Novel Advances and Approaches in Biomedical Materials Based on Calcium Phosphates

Novel Advances and Approaches in Biomedical Materials Based on Calcium Phosphates

Special Issue Editor
Michael R. Mucalo

MDPI • Basel • Beijing • Wuhan • Barcelona • Belgrade • Manchester • Tokyo • Cluj • Tianjin

Special Issue Editor
Michael R. Mucalo
University of Waikato
New Zealand

Editorial Office
MDPI
St. Alban-Anlage 66
4052 Basel, Switzerland

This is a reprint of articles from the Special Issue published online in the open access journal *Materials* (ISSN 1996-1944) (available at: https://www.mdpi.com/journal/materials/special_issues/biomaterials_phosphates).

For citation purposes, cite each article independently as indicated on the article page online and as indicated below:

LastName, A.A.; LastName, B.B.; LastName, C.C. Article Title. *Journal Name* **Year**, *Article Number*, Page Range.

ISBN 978-3-03928-264-7 (Hbk)
ISBN 978-3-03928-265-4 (PDF)

© 2020 by the authors. Articles in this book are Open Access and distributed under the Creative Commons Attribution (CC BY) license, which allows users to download, copy and build upon published articles, as long as the author and publisher are properly credited, which ensures maximum dissemination and a wider impact of our publications.

The book as a whole is distributed by MDPI under the terms and conditions of the Creative Commons license CC BY-NC-ND.

Contents

About the Special Issue Editor vii

Preface to "Novel Advances and Approaches in Biomedical Materials Based on Calcium Phosphates" ix

Michael R. Mucalo
Special Issue: Novel Advances and Approaches in Biomedical Materials Based on Calcium Phosphates
Reprinted from: *Materials* **2019**, *12*, 405, doi:10.3390/ma12030405 1

Ahmed Samir Bakry, Mona Aly Abbassy, Hanin Fahad Alharkan, Sara Basuhail, Khalil Al-Ghamdi and Robert Hill
A Novel Fluoride Containing Bioactive Glass Paste is Capable of Re-Mineralizing Early Caries Lesions
Reprinted from: *Materials* **2018**, *11*, 1636, doi:10.3390/ma11091636 7

Yunia Dwi Rakhmatia, Yasunori Ayukawa, Akihiro Furuhashi and Kiyoshi Koyano
Carbonate Apatite Containing Statin Enhances Bone Formation in Healing Incisal Extraction Sockets in Rats
Reprinted from: *Materials* **2018**, *11*, 1201, doi:10.3390/ma11071201 17

Valentina Mitran, Raluca Ion, Florin Miculescu, Madalina Georgiana Necula, Aura-Catalina Mocanu, George E. Stan, Iulian Vasile Antoniac and Anisoara Cimpean
Osteoblast Cell Response to Naturally Derived Calcium Phosphate-Based Materials
Reprinted from: *Materials* **2018**, *11*, 1097, doi:10.3390/ma11071097 33

Igor A. Khlusov, Yuri Dekhtyar, Yurii P. Sharkeev, Vladimir F. Pichugin, Marina Y. Khlusova, Nataliya Polyaka, Fjodors Tjulkins, Viktorija Vendinya, Elena V. Legostaeva, Larisa S. Litvinova, Valeria V. Shupletsova, Olga G. Khaziakhmatova, Kristina A. Yurova and Konstantin A. Prosolov
Nanoscale Electrical Potential and Roughness of a Calcium Phosphate Surface Promotes the Osteogenic Phenotype of Stromal Cells
Reprinted from: *Materials* **2018**, *11*, 978, doi:10.3390/ma11060978 49

Öznur Demir Oğuz and Duygu Ege
Rheological and Mechanical Properties of Thermoresponsive Methylcellulose/Calcium Phosphate-Based Injectable Bone Substitutes
Reprinted from: *Materials* **2018**, *11*, 604, doi:10.3390/ma11040604 75

Anne V. Boehm, Susanne Meininger, Annemarie Tesch, Uwe Gbureck and Frank A. Müller
The Mechanical Properties of Biocompatible Apatite Bone Cement Reinforced with Chemically Activated Carbon Fibers
Reprinted from: *Materials* **2018**, *11*, 192, doi:10.3390/ma11020192 93

Elisabet Farré-Guasch, Nathalie Bravenboer, Marco N. Helder, Engelbert A. J. M. Schulten, Christiaan M. ten Bruggenkate and Jenneke Klein-Nulend
Blood Vessel Formation and Bone Regeneration Potential of the Stromal Vascular Fraction Seeded on a Calcium Phosphate Scaffold in the Human Maxillary Sinus Floor Elevation Model
Reprinted from: *Materials* **2018**, *11*, 161, doi:10.3390/ma11010161 105

Eisner Salamanca, Yu-Hwa Pan, Aileen I. Tsai, Pei-Ying Lin, Ching-Kai Lin, Haw-Ming Huang, Nai-Chia Teng, Peter D. Wang and Wei-Jen Chang
Enhancement of Osteoblastic-Like Cell Activity by Glow Discharge Plasma Surface Modified Hydroxyapatite/β-Tricalcium Phosphate Bone Substitute
Reprinted from: *Materials* **2017**, *10*, 1347, doi:10.3390/ma10121347 . . . . **121**

Humair A. Siddiqui, Kim L. Pickering and Michael R. Mucalo
A Review on the Use of Hydroxyapatite- Carbonaceous Structure Composites in Bone Replacement Materials for Strengthening Purposes
Reprinted from: *Materials* **2018**, *11*, 1813, doi:10.3390/ma11101813 . . . . **133**

About the Special Issue Editor

Michael R. Mucalo received his PhD degree from the University of Auckland, New Zealand, in 1991. He has since held postdoctoral positions at what was then the Department of Scientific and Industrial Research (Chemistry Division)/Industrial Research Ltd in Lower Hutt, New Zealand, as a Foundation Post Doctoral Fellow during 1991–1993 and then as a Science and Technology Agency Fellow in the National Industrial Research Institute of Nagoya, Japan, during 1993–1995. He is currently Associate Professor in Chemistry at the School of Science, University of Waikato, Hamilton, New Zealand, where he has held an academic position since 1995 and also currently serves as Vice President of the New Zealand Institute of Chemistry. His research interests span the study and characterisation of biomedical or water treatment materials which are principally hydroxyapatite-derived or repurposed from byproducts, controlled release drug delivery, colloid science, and in situ IR spectroelectrochemistry. He has co-authored over 70 published articles in a variety of peer reviewed biomedical materials and chemistry-based journals, and currently serves on the editorial boards of three journals: *Coatings* (MDPI), *BPEX* (*Biomedical Physics & Engineering Express*—IOPscience), and *Bioactive Materials* (KeAi Publishing).

Preface to "Novel Advances and Approaches in Biomedical Materials Based on Calcium Phosphates"

The research on calcium phosphate use in the development and clinical application of biomedical materials is diverse and genuinely interdisciplinary, with much work leading to innovative solutions for improvement of health outcomes. This Special Issue aims to summarise current advances in this area. The nine published papers cover a wide spectrum of topical areas, such as (1) remineralisation pastes for decalcified teeth, (2) use of statins to enhance bone formation, (3) how dolomitic marble and seashells can be processed into bioceramic materials, (4) relationships between the roughness of calcium phosphate surfaces and surface charge with investigations into the effect on human MRC osteogenic differentiation and maturation, (5) rheological and mechanical properties of a novel injectable bone substitute, (6) improving strength of bone cements by incorporating reinforcing chemically modified fibres, (7) using adipose stem cells to stimulate osteogenesis, osteoinduction, and angiogenesis on calcium phosphates, (8) using glow discharge treatments to remove surface contaminants from biomedical materials to enhance cell attachment and improve bone generation, and (9) a review on how classically brittle hydroxyapatite based scaffolds can be improved by making fibre–hydroxyapatite composites, with detailed analysis of the mechanisms of ceramic crack propagation and their prevention via fibre incorporation into hydroxyapatite.

Michael R. Mucalo
Special Issue Editor

Editorial

Special Issue: Novel Advances and Approaches in Biomedical Materials Based on Calcium Phosphates

Michael R. Mucalo

School of Science, University of Waikato, Private Bag 3105, Hamilton 3240, New Zealand; michael.mucalo@waikato.ac.nz; Tel.: +64-7-838-4404

Received: 18 January 2019; Accepted: 24 January 2019; Published: 28 January 2019

Abstract: Research on calcium phosphate use in the development and clinical application of biomedical materials is a diverse activity and is genuinely interdisciplinary, with much work leading to innovative solutions for improvement of health outcomes. This Special Issue aimed to summarize current advances in this area. The nine papers published cover a wide spectrum of topical areas, such as (1) remineralisation pastes for decalcified teeth, (2) use of statins to enhance bone formation, (3) how dolomitic marble and seashells can be processed into bioceramic materials, (4) relationships between the roughness of calcium phosphate surfaces and surface charge with the effect on human MRC osteogenic differentiation and maturation being investigated, (5) rheological and mechanical properties of a novel injectable bone substitute, (6) improving strength of bone cements by incorporating reinforcing chemically modified fibres, (7) using adipose stem cells to stimulate osteogenesis, osteoinduction, and angiogenesis on calcium phosphates, (8) using glow discharge treatments to remove surface contaminants from biomedical materials to enhance cell attachment and improve bone generation, and (9) a review on how classically brittle hydroxyapatite based scaffolds can be improved by making fibre-hydroxyapatite composites, with detailed analysis of ceramic crack propagation mechanisms and its prevention via fibre incorporation in the hydroxyapatite.

Keywords: calcium phosphates; hydroxyapatite; scaffolds; biomedical; biocompatibility; bone substitute; bone regeneration

1. Introduction

The diversity of research on the development and use of calcium phosphate based biomedical materials gives rise to a thoroughly interesting and extremely interdisciplinary field of activity, from which many innovative advances can be realized for improving human health.

The importance of the field in today's times is of unquestionable importance. Bohner [1] has commented that bone related medical conditions and bone breaks have increased due to the progressive ageing of the global population and the popularity of high risk sporting activities. In terms of the ageing aspect, Habraken et al. [2] have cited expert opinion that in the near future, 30% of all hospital admissions will be taken up by osteoporosis patients.

In general, the main reason for pursuing calcium phosphate based bone substitutes is to reduce dependence on autologous bone, which is harvestable only in limited quantities from the iliac crest, and which necessarily subjects the patient to two surgical procedures. Donor site morbidity is also an issue. Allografts, on the other hand, are more plentiful in supply but have been said to lead to inferior healing outcomes relative to autologous bone and can potentially be a vector for disease transmission [3]. As calcium phosphate-based bone substitutes are similar in chemical composition to human bone, these have been subjected to the most research scrutiny and are relatively easy to manufacture or to source from natural sources. It is for this reason that advances in calcium phosphate based biomedical materials form the dominant subject of this Special Issue.

Numerous reviews on state-of-the-art calcium phosphate biomaterials (and indeed biomaterials in general, constructed from other non-calcium phosphate materials) have been written. Some of these have been general in nature but others have concentrated on specific aspects or materials being actively worked upon in the field. Examples include "biphasic calcium phosphates (BCPs)" which have been described as the "gold standard of bone substitutes in bone reconstructive surgery" [4]. BCPs consist of varying mixtures of hydroxyapatite (HAp) and beta tricalcium phosphate (β-TCP), with the degree of solubility of the ceramic mixture being important. The behaviour of osteoclastic resorption was found to be influenced by the HAp/ β-TCP ratio [4]. BCPs (the term being in existence since 1986) have been successfully employed for bone reconstructive purposes in a variety of clinical situations. Other biomedical materials reviews have summarized the importance of naturally derived materials, such as animal bone (cattle bone, fish bone, etc.), in providing a feedstock source for creating calcium phosphate based materials [5]. Another active area of study is bone cements [6], which are widely used in bone reconstructive surgery due to their excellent bioresorbability, which obviates the need for xenografts or allografts that may be only semi absorbed. These materials are also useful as delivery vehicles for medications, such as antibiotics used to combat infection and biofilm production that often find an ideal milieu to proliferate on implants or bone replacement materials implanted in vivo.

Research surrounding the effects of cationic or anionic substitutions into hydroxyapatite on its biofunctionality is also a vast activity, currently with much written the subject, as is evident in this recent review by Tite et al. [7], though this only concentrates on cationic substitutions. The doping of cations or anions into hydroxyapatite has been shown to have a number of beneficial effects, such as increasing biological activity in terms of cytocompatibility, cell viability and proliferation, adhesion, haemocompatibility, antimicrobial or antifungal factors, among other effects. However, the field is crowded with many different studies, and some degree of comparison between studies to ascertain the best procedures to follow is needed to drill down into what exactly is causing the beneficial effects of a cation or an anion substituted HAp lattice.

Ceramic composites are also an active area of study [8], with the intent being to more closely mimic bone, which can be described as a living composite of bone and collagen [9]. Synthetic HAp on its own is too brittle and clearly unsuitable for load-bearing applications due to the risk of cracking and failure. Composite scaffold production has the aim of producing materials with superior mechanical properties.

Other work has shown that processing conditions and controlling aspects as fundamental as the pH of the solution in which the hydroxyapatite is generated can influence biological response [10]. This can alter surface properties, such as zeta potential, which can, among other factors, dictate the success of cell interactions with the implanted biomedical material.

2. Synopsis of Papers Contributed to the Special Issue

In discussing some of the main areas of biomedical materials research involving calcium phosphate-based materials above, this Special Issue aimed to attract papers describing the latest advances, and successfully led to nine papers being accepted for publication that provide a very interesting snapshot into some very topical areas of biomedical research involving calcium phosphate-based materials. A brief summary of these is provided below and interested readers are invited to seek further information by referring to the actual articles embodied in this Special Issue.

The paper by Bakry et al. [11] contains a dental materials research theme. Orthodontic patients can frequently suffer from decalcification of enamel surfaces adjacent to their orthodontic devices due to problems with oral hygiene. These so-called white spot lesions are difficult to treat and can lead to dental caries, which can become unsightly. This paper examined the use of a novel fluoride-containing bioactive glass paste known as BiominF for remineralization of white spot lesions associated with orthodontic treatment. The study used transverse micro radiography and scanning electron microscopy/energy dispersive spectroscopy (SEM/EDS) techniques to observe remineralization events on enamel demineralized surfaces.

In the paper by Rakhmatia et al. [12], an investigation into how statins can enhance bone formation was carried out. This was executed by evaluating how readily apatite blocks made via a dissolution-reprecipitation reaction, involving preset gypsum in the presence and absence of statins, could increase bone formation in a socket healing scenario after tooth extraction. Four week old male Wistar rats were used as the model to test this hypothesis. Carbonate apatite containing statins were found to exhibit very favourable results for enhancing bone mineral density in the vertical plane showing it could not only allow but promote bone healing in socket healing. This result could, in the future, be a strategy for the preservation of alveolar bone after tooth extraction.

The research reported in this paper by Mitran et al. [13] followed the theme of deriving biomedical materials from naturally sourced feedstocks. Bioceramic materials were obtained from processing of dolomitic marble and indigenously sourced *Mytilus galloprovincialis* seashells via the decomposition of the calcium carbonate minerals contained within the natural powders into CaO, hydrolysis to form $Ca(OH)_2$, and then mixing with phosphoric acid to form pellets. These were found to contain a biphasic composition involving largely hydroxyapatite and brushite. These were then evaluated for their bone regenerative capabilities by investigating the interaction of pre-osteoblastic cells, namely MC3T3-E1 cells, with the materials in terms of their adhesion properties, morphological characteristics, viability, proliferation, and differentiation. The *Mytilus galloprovincialis* seashell-derived material was found, in particular, to efficaciously induce cell differentiation in these studies on a par with the performance of the reference hydroxyapatite materials used.

Khlusov et al.'s paper [14] detailed the very interesting area of surface electrical charge and topography of biomedical materials and their influence on cell responses to these materials.

In this in vitro study, the authors investigated the relationship between the roughness of calcium phosphate surfaces and the surface electrical charge, and its possible role in affecting human MSC (mesenchymal stem cells) osteogenic differentiation and maturation. The behaviour of these in vitro was estimated by human adipose-derived MSCs or prenatal stromal cells from the human lung. A microarc calcium phosphate coating was prepared on a titanium substrate and characterized, although was found to be amorphous to X-ray in its as-prepared form, hence necessitating its annealing at 1073 K to crystallize the phases present to enable identification. These were revealed to be a mixture of calcium titanium phosphates, calcium/titanium pyrophosphates, and anatase. The authors then used human adult adipose derived MSCs or human lung derived prenatal stromal cells (abbreviated in the paper as "HLPSCs") and cultured them on the calcium phosphate surfaces to estimate MSC behaviour. It was demonstrated that roughness, non-uniform charge polarity, and the electrical potential of the microarc coatings affected the osteogenic differentiation and maturation of the cells in vitro. Electrical potential on the surface of the coatings increased with increasing roughness at the surface (at microscale). It was found that nanoscale surface features influenced the sign of the electrical potential with location of negative electrical charges mostly existing in micro and nanosockets of the coating surface. Positive charges mostly existed at nanorelief peaks. In the sockets, HLPSCs were located and expressed osteoblastic markers osteocalcin and alkaline phosphatase. The topography of the coating promoted the osteoblast phenotype of HLPSCs. Overall, it was thought that the negative sign existing at the calcium phosphate surfaces and its magnitude at various locations corresponding to micro and nanosockets could be factors that stimulate osteoblastic differentiation and maturation of human stromal cells.

Rheological and mechanical properties of a novel injectable bone substitute are the theme of the paper by Oğuz et al. [15]. This was accomplished by mixing a bioceramic powder in a solution of polymers consisting of methylcellulose, gelatin, and citric acid. Methylcellulose was used because of its thermoresponsive and biocompatibility properties. Added gelatin and citric acid served to adjust rheological properties in the injectable bone substitute. The bioceramic powder component comprised tetracalcium phosphate, hydroxyapatite, brushite, and calcium sulfate dihydrate, which was added in proportions up to 50 wt%. The so-prepared injectable bone substitute had a "chewing gum" consistency. Characterisation was carried out on the materials in terms of chemical structure, rheological, and mechanical properties. Hardening of the injectable material was confirmed at physiological temperature. In addition, in vitro degradation studies were carried out. These indicated a lower rate of degradation as the wt% of bioceramic powder increased, with a concomitant improvement in mechanical properties. Overall, the injectable bone substitute material was judged a promising candidate for treatment of bone defects in non-load bearing applications.

In yet another bone-cement-themed paper, this time by Boehm et al. [16], the authors considered solutions to the long term and well known problem of the brittleness of calcium phosphate cements, which leads to low fracture toughness. This, of course, limits them to non-load bearing applications in clinical practice. Carbon fibre reinforcement is known to improve fracture resistance, but at the same time compromise strength of the composite. It was found that chemically modifying the fibre surfaces and using these fibres to reinforce the calcium phosphate cements resulted in a modification of the fibre matrix interface and the fracture behaviour. In taking this approach, the authors were able to demonstrate enhanced mechanical properties as regards bending strength and work of fracture to a strain of 5%. They found that using fibre reinforcement did not affect the cell proliferation and activity of MG63 (human osteoblast-like) cells that were used as biocompatibility markers. The conclusion was that the use of chemically activated C-fibres to reinforce calcium phosphate cements was a promising method to achieve better mechanical properties in these bone cement replacement materials to the extent that they might be of use in load bearing applications.

Farré-Guasch et al.'s paper [17] tries to address the classical issue encountered with xenograft type materials in regard to their lack of osteoinductive and angiogenic properties. This paper concentrated on the potential of adding adipose stem cells (ASC) to calcium phosphate based scaffolds in order to stimulate osteogenesis, osteoinduction, and angiogenesis in them. The bone substitute ("calcium phosphate carrier") materials used for seeding the (autologous) ASC-containing stromal vascular fraction from ten patients were (pure) β-tricalcium phosphate and biphasic calcium phosphate. These were used in a human maxillary sinus floor elevation model in a one-step surgical procedure. After six months of implantation, biopsies collected quantified blood vessel formation and bone percentages. Bone percentages were found to correlate with blood vessel formation and were higher in study subjects compared to control biopsies in the cranial area. This was particularly so with the β-tricalcium phosphate blocks. The study showed the pro-angiogenic effect of using the stromal vascular fraction to seed the calcium phosphate scaffolds.

The paper by Salamanca et al. [18] is concerned with surface treatments for the purpose of removing contaminants on biomedical material surfaces. To achieve this, glow discharge plasma treatments on calcium phosphate biomaterial surfaces were carried out. The point of removing contaminants was in order to facilitate cell attachment and enhance bone regeneration. These contaminants can arise from inorganic grit blasting media (which can be Si or Al-based oxides) residues or layers arising from processing fluids, such as etchants or cleaning solvents, or substances that remain after sterilization procedures, like autoclaving or the use of ethylene oxide. In addition, most surfaces exposed to ambient air can pick up organic contamination (often known as "adventitious hydrocarbons" in surface science [19]). In this study, calcium phosphate composite materials consisting of hydroxyapatite and β-tricalcium phosphate were treated with argon glow discharge plasma for 15 min at room temperature and were then subjected to scanning electron microscopy/energy dispersive spectroscopic (SEM/EDS) analysis. These showed that low levels of metal ion impurities at the surface had been removed by the glow discharge process. Other biological tests, such as cell viability, morphology, and an alkaline phosphatase assay, were also conducted on untreated and glow discharge treated calcium phosphate surfaces and compared to controls. Improved cell proliferation, increased alkaline phosphatase activity, and enhanced differentiation into osteoblast like cells were observed after 5 days on the argon glow discharge treated substrates. No chemical modification was noted to occur to the bulk calcium phosphate materials. Thus, the improvement was directly ascribed to the argon glow discharge treatment of the surface, though it was recommended that further testing be conducted in the in vivo models to further prove its efficacy.

The review by Siddiqui et al. [20] discusses the use of hydroxyapatite as a bone-substituting and osteoconductive scaffold, but also comments on the classical problem of implants made from (pure) hydroxyapatite having poor mechanical properties when placed under load. This limits their usage to non-load bearing applications. In order to make hydroxyapatite more suitable for load bearing applications or to achieve better mechanical properties, it is necessary to create composites of hydroxyapatite with other materials, whilst maintaining the biocompatibility of the resultant material. The review then moves on to discuss the strengthening and toughening mechanisms which are crucial to understand if the aim is to produce new and optimized materials from the compositing approach. The main engineering conundrum when designing a material that is both strong and tough is that often an increase in strength can be coupled with a decrease in toughness, and vice versa. A detailed discussion is provided on fracture mechanics, which constitutes a valuable tool to analyze the conditions under which a crack may propagate and eventually lead to failure. Further discussion concentrates on how crack propagation can be prevented, which in brittle ceramic materials is known as a process of extrinsic toughening.

3. Concluding Remarks

As is obvious, a diverse range of topics has been covered in this Special Issue, although it is clear that there is an underlying emphasis on how the biological processes, such as cell proliferation and others, can be facilitated or enhanced on the prepared calcium phosphate biomaterials when implanted. This remains of critical importance when engaging in this research area. It is hoped that the readers will find this collection of research of interesting reading for advancing understanding in calcium phosphate-based biomaterials.

Conflicts of Interest: The author declares no conflict of interest.

References

1. Bohner, M. Resorbable biomaterials as bone graft substitutes. *Mater. Today* **2010**, *13*, 24–30. [CrossRef]
2. Habraken, W.; Habibovic, P.; Epple, M.; Bohner, M. Calcium phosphates in biomedical applications: materials for the future? *Mater. Today* **2016**, *19*, 69–87. [CrossRef]
3. Wang, W.; Yeung, K.W.K. Bone grafts and biomaterials substitutes for bone defect repair: A review. *Bioactive Mater.* **2017**, *2*, 224–247. [CrossRef] [PubMed]
4. Bouler, J.M.; Pilet, P.; Gauthier, O.; Verron, E. Biphasic calcium phosphate ceramics for bone reconstruction: A review of biological response. *Acta Biomater.* **2017**, *53*, 1–12. [CrossRef] [PubMed]
5. Mucalo, M.R. Animal-bone derived hydroxyapatite in biomedical applications. In *Hydroxyapatite (HAp) for Biomedical Applications*; Elsevier: New York, NY, USA, 2015; pp. 307–342.
6. Van Staden, A.D.; Dicks, L.M.T. Calcium orthophosphate-based bone cements (CPCs): applications, antibiotic release and alternatives to antibiotics. *J. Appl. Biomater. Funct. Mater.* **2012**, *10*, 2–11. [CrossRef] [PubMed]
7. Tite, T.; Popa, A.-C.; Balescu, L.M.; Bogdan, I.M.; Pasuk, I.; Ferreira, J.M.F.; Stan, G.E. Cationic substitutions in hydroxyapatite: current status of the derived biofunctional effects and their in vitro interrogation methods. *Materials* **2018**, *11*, 2081. [CrossRef] [PubMed]
8. Turnbull, G.; Clarke, J.; Picard, F.; Riches, P.; Jia, L.; Han, F.; Li, B.; Shu, W. 3D bioactive composite scaffolds for bone tissue engineering. *Bioactive Mater.* **2018**, *3*, 278–314. [CrossRef] [PubMed]
9. Simkiss, K. Bone and biomineralization. *Inst. Biol. Stud. Biol.* **1975**, *53*, 60.
10. Cox, S.C.; Parastoo, J.; Williams, R.; Grover, L.; Mallick, K.K. The importance of processing conditions on the biological response to apatites. *Powder Technol.* **2015**, *284*, 195–203. [CrossRef]
11. Bakry, A.S.; Abbassy, M.A.; Alharkan, H.F.; Basuhail, S.; Al-Ghamdi, K.; Hill, R. A Novel Fluoride Containing Bioactive Glass Paste is Capable of Re-Mineralizing Early Caries Lesions. *Materials* **2018**, *11*, 1636. [CrossRef] [PubMed]
12. Rakhmatia, Y.D.; Ayukawa, Y.; Furuhashi, A.; Koyano, K. Carbonate Apatite Containing Statin Enhances Bone Formation in Healing Incisal Extraction Sockets in Rats. *Materials* **2018**, *11*, 1201. [CrossRef] [PubMed]

13. Mitran, V.; Ion, R.; Miculescu, F.; Necula, M.G.; Mocanu, A.-C.; Stan, G.E.; Antoniac, I.V.; Cimpean, A. Osteoblast Cell Response to Naturally Derived Calcium Phosphate-Based Materials. *Materials* **2018**, *11*, 1097. [CrossRef] [PubMed]
14. Khlusov, I.A.; Dekhtyar, Y.; Sharkeev, Y.P.; Pichugin, V.F.; Khlusova, M.Y.; Polyaka, N.; Tyulkin, F.; Vendinya, V.; Legostaeva, E.V.; Litvinova, L.S.; et al. Nanoscale Electrical Potential and Roughness of a Calcium Phosphate Surface Promotes the Osteogenic Phenotype of Stromal Cells. *Materials* **2018**, *11*, 978. [CrossRef] [PubMed]
15. Demir Oğuz, Ö.; Ege, D. Rheological and Mechanical Properties of Thermoresponsive Methylcellulose/Calcium Phosphate-Based Injectable Bone Substitutes. *Materials* **2018**, *11*, 604. [CrossRef] [PubMed]
16. Boehm, A.V.; Meininger, S.; Tesch, A.; Gbureck, U.; Müller, F.A. The Mechanical Properties of Biocompatible Apatite Bone Cement Reinforced with Chemically Activated Carbon Fibers. *Materials* **2018**, *11*, 192. [CrossRef] [PubMed]
17. Farré-Guasch, E.; Bravenboer, N.; Helder, M.N.; Schulten, E.A.J.M.; Ten Bruggenkate, C.M.; Klein-Nulend, J. Blood Vessel Formation and Bone Regeneration Potential of the Stromal Vascular Fraction Seeded on a Calcium Phosphate Scaffold in the Human Maxillary Sinus Floor Elevation Model. *Materials* **2018**, *11*, 161. [CrossRef] [PubMed]
18. Salamanca, E.; Pan, Y.-H.; Tsai, A.I.; Lin, P.-Y.; Lin, C.-K.; Huang, H.-M.; Teng, N.-C.; Wang, P.D.; Chang, W.-J. Enhancement of Osteoblastic-Like Cell Activity by Glow Discharge Plasma Surface Modified Hydroxyapatite/β-Tricalcium Phosphate Bone Substitute. *Materials* **2017**, *10*, 1347. [CrossRef] [PubMed]
19. Barr, T.L.; Seal, S. Nature of the use of adventitious carbon as a binding energy standard. *J. Vac. Sci. Technol. A* **1995**, *13*, 1239–1246. [CrossRef]
20. Siddiqui, H.A.; Pickering, K.L.; Mucalo, M.R. A Review on the Use of Hydroxyapatite-Carbonaceous Structure Composites in Bone Replacement Materials for Strengthening Purposes. *Materials* **2018**, *11*, 1813. [CrossRef] [PubMed]

© 2019 by the author. Licensee MDPI, Basel, Switzerland. This article is an open access article distributed under the terms and conditions of the Creative Commons Attribution (CC BY) license (http://creativecommons.org/licenses/by/4.0/).

Article

A Novel Fluoride Containing Bioactive Glass Paste is Capable of Re-Mineralizing Early Caries Lesions

Ahmed Samir Bakry [1,2,*], Mona Aly Abbassy [3,4], Hanin Fahad Alharkan [3], Sara Basuhail [3], Khalil Al-Ghamdi [5] and Robert Hill [6]

1. Operative Dentistry Department, Faculty of Dentistry, King Abdulaziz University, Jeddah 21589, Saudi Arabia
2. Conservative Dentistry Department, Faculty of Dentistry, Alexandria University, Alexandria 21614, Egypt
3. Department of Orthodontics, Faculty of Dentistry, King Abdulaziz University, Jeddah 21589, Saudi Arabia; monaabbassy@gmail.com (M.A.A.); haneenharkan@gmail.com (H.F.A.); sara.basuhail@hotmail.com (S.B.)
4. Dental Department, Alexandria University, Alexandria 21614, Egypt
5. Primary Health care Centre, Sabt Alalayah Hospital, Ministry of Health, Bisha 67611, Saudi Arabia; khlilsaeed@gmail.com
6. Physical Sciences in Relation to Dentistry, Institute of Dentistry, Dental Physical Sciences Unit, Queen Mary University of London, London E1 2AD, UK; r.hill@qmul.ac.uk
* Correspondence: drahmedbakry@gmail.com; Tel.: +966-(2)-6400-000; Fax: +966-(12)-695-2437

Received: 1 July 2018; Accepted: 23 August 2018; Published: 6 September 2018

Abstract: White-spot-lesions (WSL) are a common complication associated with orthodontic treatment. In the current study, the remineralization efficacy of a BiominF® paste was compared to the efficacy of a fluoride gel. Methods: Orthodontic brackets were bonded to 60 human premolars buccal surfaces, which were covered with varnish, except a small treatment area (3 mm^2). All specimens were challenged by a demineralization solution for 4 days. Specimens were assigned into 4 groups: BiominF® paste, Fluoride (4-min application), fluoride (twenty four hours application), and the control (n = 15). After cross-sectioning, enamel slabs having a thickness of approximately 100–120 µm were obtained. A TMR (Transverse Micro Radiography) technique was used to observe the sub-surface enamel lesions' depth and mineral density, and their response to the remineralization protocols. One way ANOVA was used to analyze the results (α = 0.05). The top and the cross-sectional surfaces were observed using SEM/EDS. Results: Specimens treated with BiominF® paste showed significant decrease in delta z values, however lesion depth showed no significant difference when compared to the other three groups (p < 0.05). SEM/EDS observation showed the formation of crystal like structures on top of enamel demineralized surfaces, when treated with BiominF® paste. In conclusion BiominF® paste can be considered an effective remineralizing agent for white spot Lesions.

Keywords: fluoride bioactive glass; biomin; caries; enamel

1. Introduction

Many orthodontic patients seek improvement in their dental aesthetic features; however orthodontic treatment is often associated with the development of white spot lesions, which is extremely difficult to treat [1]. Moreover, progress of these lesions to form cavities occurs frequently in many cases, thus jeopardizing the final expected esthetic results [1]. The increase in caries risk factors in orthodontic patients may be attributed to the presence of many retentive areas surrounding the enamel brackets, which retain large quantities of cariogenic oral biofilm [2]. Previous studies showed that the levels of cariogenic biofilm in the oral cavities of orthodontic patients, may be 2–3 times higher than in normal individuals suffering from high rates of biofilm formation [3].

Moreover, the spread of dietary habits and assorted systemic [4–6] and genetic disorders [7], may lead to the development of various early enamel lesions, which need re-mineralizing [8] rather than surgically treating these lesions through composite restorations. The need for development of efficient remineralizing agents, encouraged researchers to investigate the possibility of using bioactive glasses, for the re-mineralization of enamel [9,10] and dentin lesions [11,12].

Results reporting the efficacy of using bioactive glasses in treating dental lesions, have suggested that adding small amounts of calcium fluoride to a bioactive glass may enhance the formation of fluoroapatite, which is expected to exhibit acid resistance to acidic cariogenic attacks [13].

In the current study, the trans-microradiography technique was used together with the SEM/EDS technique [9,10] to investigate the capability of the BiominF® paste as a potent remineralizing agent.

The hypothesis adopted in the current study was that the BiominF® paste will be able to re-mineralize the artificial white spot lesion induced on the enamel surfaces.

2. Materials and Methods

2.1. Specimens' Preparation

The teeth's buccal surfaces were cleaned thoroughly using pumice. Unitek™ Etching Gel (3M Unitek, Monrovia, CA, USA) was applied exclusively to the enamel surfaces, onto which the orthodontic brackets were bonded for 15 s, followed by rinsing using distilled water and drying by air-way syringe for 15 s. The orthodontic metallic brackets (MiniSprint®, Forestadent, Pforzheim, Germany) were carefully bonded to the buccal surfaces of all specimens, without extruding any of the Transbond XT primer or the Transbond PLUS color change adhesive (3M Unitek) to the area of observation, next to the cemented Orthodontic brackets [1]. All specimens were embedded in resin material. Preparation and examination of the specimens are summarized in (Figure 1).

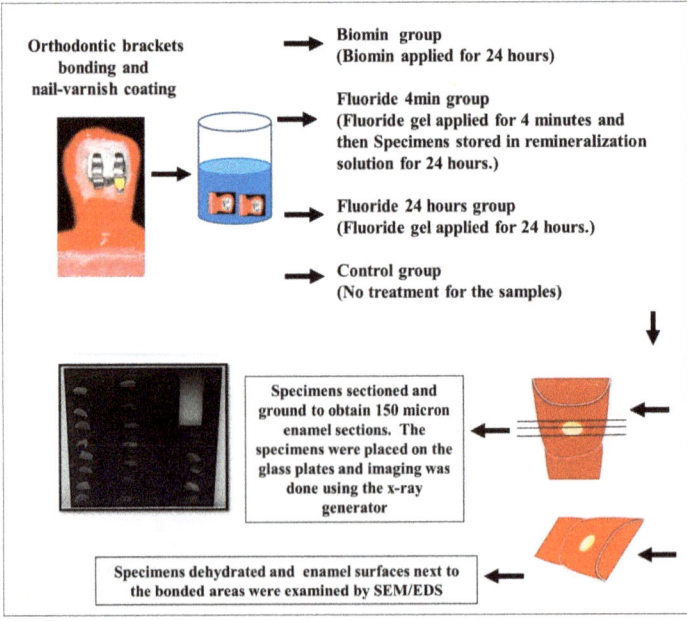

Figure 1. Samples preparation and examination.

The nail-varnished specimens were stirred in the demineralization solution (2.2 mM $CaCl_2$, 10 mM NaH_2PO_4, 50 mM acetic acid, 100 mM NaCl, 1 ppm NaF, 0.02% NaN_3; pH 4.5), for 4 days using a low-speed (100 rpm) magnetic stirrer [10]. The demineralized specimens were assigned into 4 groups ($n = 25$). There were 25 specimens that had a fluoride gel (1.23% acidulated-phosphate-fluoride, Gelato Gel, NJ, USA) (fluoride group) applied for 4 min and then gently wiped by a moist gauze; 25 specimens had the same fluoride gel applied and was not washed away (Fluoride-24 h), but rather covered by a protective light-cured resin material layer for 24 h; 25 specimens had BioMinF® applied on its surface (Biomin group); while the rest of the specimens received no further treatment (control group). All specimens were stored in a remineralizing solution (1.0 mM $CaCl_2$, 3.0 mM KH_2PO_4, 100 mM acetate, 100 mM NaCl, 0.02%, NaN_3; pH 6.3) according to Reference [6], for 24 h [10].

2.2. Biomin Application

One tenth of a gram of BioMinF® powder composed of (22–24 mol % Na_2O, 28–30 mol % CaO, 4–6 mol % P_2O_5, 36–40 mol % SiO_2, and 1.5–3.0 mol % CaF_2) was mixed on a glass slab with 2 drops of 50 wt% phosphoric acid, which was prepared by the diluting 85 wt% phosphoric acid (Wako, Osaka, Japan). The resulting paste had a pH 2.5. The BioMinF® paste was applied onto the enamel surfaces of the (Biomin group) by microbrush. The aforementioned method of application was utilized previously, for the application of another bioactive material [1,9–12,14].

2.3. Application of Bonding Agent

All specimens in the BioMinF® or Fluoride-24 h groups had their remineralizing agents protected by a layer of bonding agent (Clearfil SE Bond, Kuraray-Medical, Tokyo, Japan), which was applied over these remineralizing agents then light-cured [1,9–12,15–18] using LED (Woodpecker™ LED Curing light, China with an output of 900 mW/cm^2). After storage in the re-mineralizing solution for 24 h, the temporary bonding agent layer was removed carefully by a sharp instrument, as was previously described in References [1,9–12,15–18].

2.4. TMR Analysis

Fifteen specimens from each group were assigned for the TMR analysis. Preparation of the specimens started by dehydrating them in ascending alcohol solutions, followed by immersing the specimens in styrene monomer, and then finally embedding the specimens in low-viscosity polyester resin (Rigolac, Oken, Tokyo, Japan) in a special mold. The aforementioned low-viscosity resin, has the ability to penetrate the porous enamel surface resulting from the demineralization process. The resin penetration into the enamel surface would help in reinforcing the demineralized enamel surface and decrease the incidence of sample cracking to be 10% of the total utilized samples. After the polymerization of the embedding resin, the specimens were cut using a low-speed diamond saw (Isomet; Buehler, IL, USA) and were ground by SiC abrasive papers having different grits (ranging from 800 till 1200 grit) to obtain sections, which were approximately 100 to 120 µm in thickness [14,19–21]. The x-ray generator (CMR 2; Softex, Tokyo, Japan) was adjusted to generate 20 kV voltage and 2 mA currents for 10 min. All specimens were placed on a sensitive X-ray glass plate film (High Precision Photo-Plate-PXHW, Konica, Tokyo, Japan). The TMR images, together with 15 aluminum step-wedges, were captured in the sensitive glass plate film, and were digitized using a digital camera attached to a microscope (ML 8500, Meiji, Techno, Japan). The relative mineral density (%) was calculated as was previously reported in References [14,19–21]. The definitions of lesion depth and the mineral loss (ΔZ, vol % µm), followed the previously published data in Reference [14,19–21].

2.5. SEM/EDS Top Surface Examination

The white spot lesions that were formed in five specimens from each group, were examined by SEM/EDS (JCM-6000 NeoScope, JEOL, Tokyo, Japan). All specimens were dehydrated followed by gold coating. The specimens' surface and chemical composition were observed by SEM/EDS (JCM-6000 NeoScope, JEOL), for the following elements phosphorus, calcium, oxygen, fluoride, carbon, oxygen, and silicon.

2.6. SEM/EDS Interface Preparation

Five specimens from each group were embedded in light cured resin system, according to the manufacturer's instructions. The specimens were then cross-sectioned perpendicular to the interface to give 2-mm-thick slabs. The transverse sectioned surfaces were ground and then polished with diamond pastes, down to 0.25 µm. The specimens were dehydrated then gold coated [13]. The cross-sections of the interface were examined using the SEM/EDS (JCM-6000 NeoScope, JEOL), according to Reference [22]. Line scans were done using the EDS attachment, across the treated enamel surfaces to detect the following elements: phosphorus, calcium, oxygen, fluoride, carbon, oxygen, and silicon.

2.7. Statistical Analysis

The results of the mineral loss (ΔZ, vol % µm) and lesions depths of all groups were analyzed using one-way ANOVA, followed by a Tukey test ($p < 0.05$). (The software used was SPSS 10.0 (SPSS Inc., Chicago, IL, USA)).

3. Results

3.1. Transverse Microradiography

The TMR images obtained for specimens of the 4 groups, are shown in Figure 2. The means and standard deviations of ΔZ, and lesion depth, are shown in Figures 3 and 4. The Biomin group showed statistically significant reductions in ΔZ values, but there was no significant reduction in lesion depth values for the Biomin group, when compared to the rest of specimens in the remaining groups ($p < 0.05$).

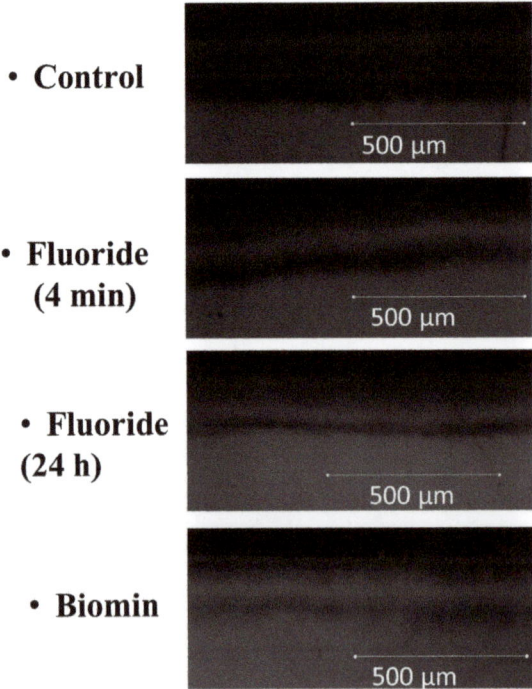

Figure 2. Transverse Micro Radiography (TMR) images for the four groups.

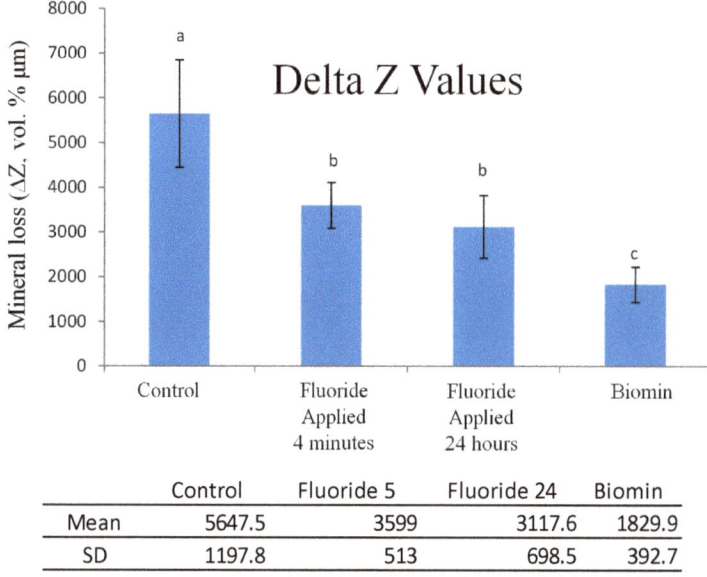

Figure 3. Mineral loss (ΔZ) after treatment protocols presented by bar chart and Table. Bars with same letters were not statistically significant α = 0.05.

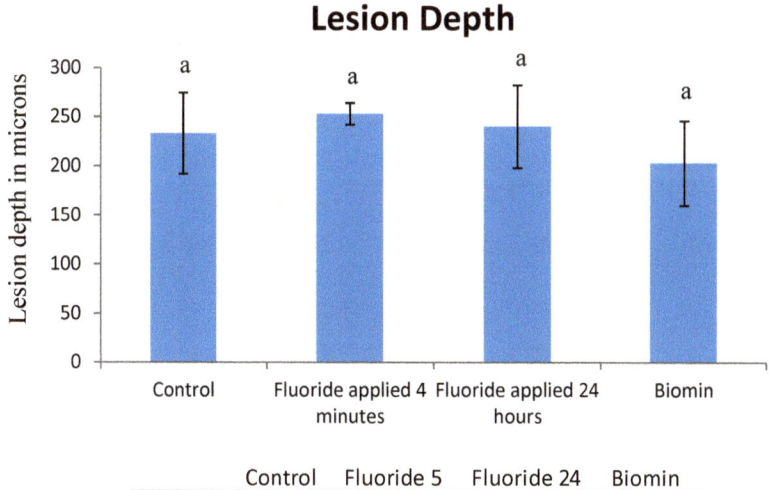

Figure 4. The results of Lesion Depth in μm, presented by bar chart and table, showing significant decrease in lesion depth for the Biomin group, compared to the remaining groups. A = 0.05. Bars with same letters were not statistically significant α = 0.05.

3.2. SEM/EDS Top Surface Examination

The (Control) group, the (fluoride-4 min), and the (Fluoride-24 h) groups' examination showed signs of enamel demineralization, with obvious boundaries of the enamel prisms (Figure 5A–C);

indicating the weak remineralizing potential of the fluoride containing agents with different application modes, utilized in the current study. EDS results showed decreased values for the mass% of the phosphorus and calcium. On the other hand, the Biomin group showed the deposition of crystal like structures, covering the whole de-mineralized enamel surface (Figure 5D). EDS analysis showed that the aforementioned layer was rich in calcium and phosphate contents, with trace amounts of silica.

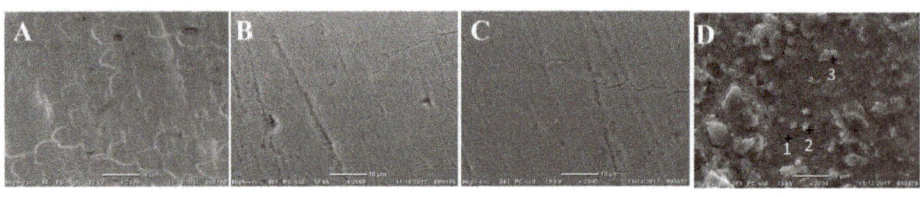

	Control	Fluoride 4 Minutes	Fluoride 24 hours	Biomin (D1)	Biomin (D2)	Biomin (D3)
Carbon	25.91	25.89	24.91	15.39	13.2	21.08
oxygen	53	53.02	54	43.1	51.46	50.48
phosphate	13.04	13	14.04	17.46	14.77	8.13
Calcium	8.06	8.1	7.06	21.58	19.21	6.99
Silica	0	0	0	1.76	0.57	12.84

Figure 5. Top surface examination by SEM/EDS for the four groups. (Control) group (**A**), the (fluoride 4 min) (**B**), and the (Fluoride-24 h) (**C**) show the boundaries of the enamel prisms. Biomin group (**D**) shows the formation of a newly formed layer having crystal like structures and silica particles. The table shows the mass percentages of each element detected by EDS in each group.

3.3. SEM/EDS Interface Examination

Specimens of the control group, the fluoride 4 min group, and the 24 h fluoride showed a decrease in the calcium and phosphorus peaks, as line scans crossed from the enamel to the embedding resin material (Figure 6A–C). The outer most enamel surface showed some weak peaks of calcium and phosphate however; these peaks were extremely weak when compared with the peaks of calcium and phosphate of normal enamel. Moreover, the subsurface enamel lesion topographical features showed roughness of the surface, indicating the presence of multiple voids and defects created between the enamel crystals in the demineralized enamel lesion.

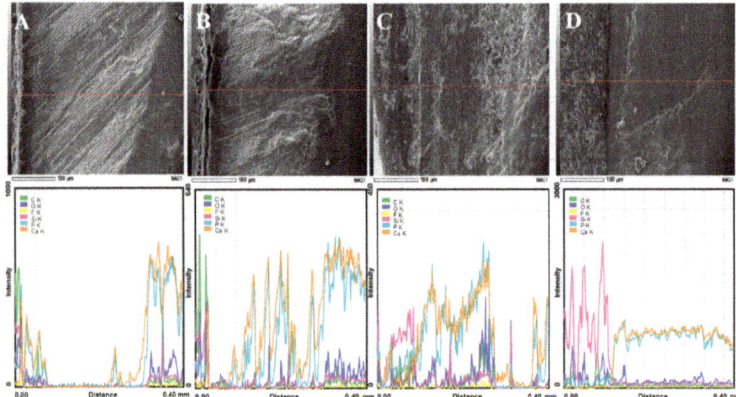

Figure 6. The (Control) group (**A**), the (fluoride 4 min) (**B**), and the (Fluoride-24 h) (**C**) specimens showed demineralization of the subsurface area, with deterioration of the micro-morphological features of the demineralized areas. Biomin group (**D**) showed remineralization of the subsurface areas, with strong calcium and phosphate peaks in the surface re-mineralized areas.

Specimens of Biomin group (Figures 6D and 7) showed the formation of a calcium phosphate rich layer on top of the enamel surface, and the subsurface enamel area exhibited a smooth area.

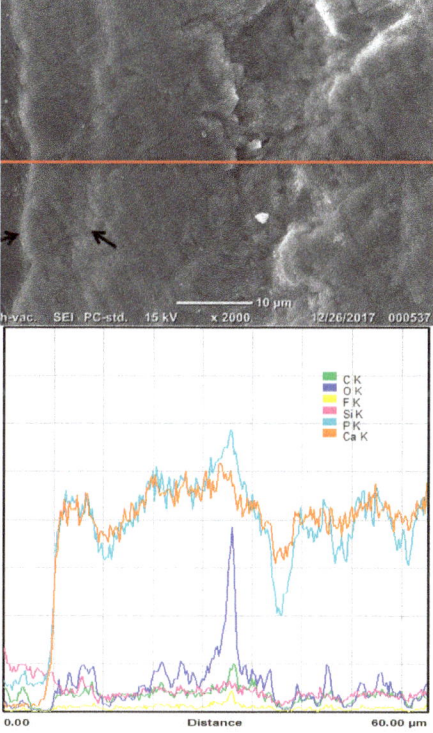

Figure 7. Examination of the Biomin group (D) with high magnification. Pointers are pointing to the newly formed layer on the demineralized enamel with strong calcium and phosphate peaks.

4. Discussion

The null hypothesis adopted in the current study was accepted. Previous research [22], showed the difficulty of re-mineralizing the sub surface demineralized enamel lesions, due to the difficulty of localizing significant high concentration of calcium, phosphate, and fluoride ions to promote the effective enamel subsurface remineralization. However, in the current experiment, biomin was capable of remineralizing the sub-surface enamel lesion efficiently, as was demonstrated by the results of the TMR experiment. TMR study is considered the gold standard for conducting experiments for testing the remineralization capacity of any agent, as described in References [19,20,22,23].

To test the remineralization efficacy of Biomin to treat the sub-surface enamel lesion, we compared its efficacy to the application of an agent containing a high concentration of fluoride (9000 ppm fluoride). Previous research [24,25], showed that agents containing high concentration of fluoride are associated with better remineralization to enamel lesions. The approximate lesion depth of the demineralized subsurface lesion induced in the current study exceeded 200 microns, previous studies showed that subsurface enamel lesions of about 100 microns needed approximately 28 days to re-mineralize in artificial saliva, without being exposed to any demineralization challenges [26]. This may explain the poor remineralization capacity of the fluoride agents utilized in the current study, which was observed by TMR. The poor fluoride remineralization pattern detected in the current study may be attributed to; firstly, the short application period, i.e., 24 h or 4 min; and secondly, the high affinity of the fluoride ions to form fluoroapatite when combined with hydroxyapatite crystals present in the superficial layers of the demineralized enamel. This might have decreased the penetration of the fluoride ions into the deeper layers of the enamel sub-surface lesion [9,23].

On the other hand, the low fluoride content of the BioMinF® (in contrast to the high fluoride content of the utilized fluoride gel), allowed the penetration of BioMinF® rich content of calcium and phosphate through the porous enamel sub-surface, causing the re-mineralization of the demineralized enamel lesion rather than remineralizing the outer enamel surface, as was observed when using fluoride as a sole remineralizing agent. Moreover, previous research showed that the presence of low concentration of fluoride in the BioMinF® glass facilitated apatite formation, since fluorapatite forms at about a unit of pH lower than hydroxyapatite, and fluoride is known to catalyze the conversion of brushite, octacalcium phosphate, and amorphous calcium phosphate to apatite [27]; which may explain the decrease in the roughness of the BioMinF® glass specimens' SEM-EDS images, due to the repair action exerted by the BioMinF® paste [13,27–29].

Based on the obtained results from the TMR and the SEM/EDS examination, it is suggested that the mechanism of forming the apatite layer after the BioMinF® may be summarized as follows: the BioMinF® powder that was mixed with 50% phosphoric acid released calcium, phosphate, fluoride, and sodium ions onto the demineralized enamel surface [9,20,21], causing the mobilization of some calcium and phosphate ions from the enamel surface [1,5–8,23]. The phosphate ions released from BioMinF® powder and the diluted phosphoric acid solution, in addition to the calcium ions released from BioMinF® were preserved from being diluted in the storage media (Saliva in case of clinical application) by the action of the protective bonding agent layer. The high concentration of calcium and phosphate ions will tend to move to the deep areas of the sub-surface enamel lesion, rendering this sub-surface lesion saturated with calcium and phosphate ions capable of repairing the voids and defects resulting from the acidic challenge. Moreover, some calcium and phosphate ions will be catalyzed by the presence of the fluoride ions to form a layer of acidic calcium-phosphate salts, on top of the demineralized enamel surface (as was observed by the SEM/EDS examination) [9,19,20]. The detection of trace amounts of silica on top of the enamel surface after Biomin paste application, may be attributed to the breaking down of the silica network of the BioMinF® by the action of the water content of the aqueous part of the acidic solution used to mix the BioMinF® powder; it was previously reported that the silica network will form the highly soluble Si-OH groups (Silanol groups) [9,20,21], upon being mixed with water, and thus it was expected that most of these silanol containing compounds would be washed out upon being rinsed with water spray after 24 h. The difficulty in detecting the fluoride peaks may be attributed to the low photon energy of the fluoride, which may render it difficult to be detected by SEM-EDS.

The addition of a rich source of phosphate (supplied from the phosphoric acid aqueous solution) to the BioMinF powder, enhanced the formation of a significant layer of a calcium phosphate compound on top of the demineralized enamel, as was previously suggested in Reference [20]. The temporary coverage of the BioMinF paste, by a thin layer of bonding agent for 24 h, protected the calcium and the phosphate ions released by the Biomin paste from being washed out by the storage media, and provided a large reservoir of calcium–phosphate ions to be released into the demineralized enamel lesion [5,6,8,23,30].

The aforementioned hypothesis for the BiominF paste action, was based on the observation of the current study and previous research. It is suggested that further research is needed to confirm the formation of fluoroapatite, and to detect the exact role of fluoride ion concentration in the BiominF paste.

5. Conclusions

The application of the BioMinF® paste using the current technique, formed a layer rich in calcium and phosphate on top of the enamel surface, suggesting that the aforementioned layer provided the source of calcium and phosphate ions which re-mineralized the subsurface enamel surface. Storing the samples for 24 h in intimate contact with fluoride 9000 ppm, did not cause significant re-mineralization of the tested sub-surface enamel lesion. The physical properties of the white BioMinF® paste on demineralized enamel using the current technique may be acceptable to patients, because it resembles the application of temporary filling materials. However, clinical studies are needed to confirm this hypothesis and to confirm the clinical remineralization capacity of BioMinF® paste.

Author Contributions: Conceptualization, A.S.B., M.A.A. and K.A.-G. and R.H.; Methodology, A.S.B., K.A.-G., H.F.A. and S.B.; Software, A.S.B.; Validation, A.S.B. and M.A.A.; Formal Analysis, A.S.B., M.A.A. and K.A.-G.; Investigation, A.S.B. and M.A.A.; Writing-Original Draft Preparation A.S.B. and M.A.A.; Writing-Review & Editing, A.S.B. and R.H.; Supervision, A.S.B. and M.A.A.

Funding: This work was supported by the Deanship of Scientific Research (DSR), King Abdulaziz University, Jeddah, Under grant No. (D-046-165-1438). The authors, therefore, gratefully acknowledge the DSR technical and financial support.

Conflicts of Interest: The authors declare no conflict of interest.

References

1. Bakhsh, T.A.; Bakry, A.S.; Mandurah, M.M.; Abbassy, M.A. Novel evaluation and treatment techniques for white spot lesions. An in vitro study. *Orthod. Craniofac. Res.* **2017**, *20*, 170–176. [CrossRef] [PubMed]
2. Al-Bazi, S.M.; Abbassy, M.A.; Bakry, A.S.; Merdad, L.A.; Hassan, A.H. Effects of chlorhexidine (gel) application on bacterial levels and orthodontic brackets during orthodontic treatment. *J. Oral Sci.* **2016**, *58*, 35–42. [CrossRef] [PubMed]
3. Klukowska, M.; Bader, A.; Erbe, C.; Bellamy, P.; White, D.J.; Anastasia, M.K.; Wehrbein, H. Plaque levels of patients with fixed orthodontic appliances measured by digital plaque image analysis. *Am. J. Orthod. Dentofac. Orthop.* **2011**, *139*, e463–e470. [CrossRef] [PubMed]
4. Crombie, F.A.; Manton, D.J.; Palamara, J.E.; Zalizniak, I.; Cochrane, N.J.; Reynolds, E.C. Characterisation of developmentally hypomineralised human enamel. *J. Dent.* **2013**, *41*, 611–618. [CrossRef] [PubMed]
5. Caruso, S.; Bernardi, S.; Pasini, M.; Giuca, M.R.; Docimo, R.; Continenza, M.A.; Gatto, R. The process of mineralisation in the development of human tooth. *Eur. J. Paediatr. Dent.* **2016**, *17*, 322–326. [PubMed]
6. Pasini, M.; Giuca, M.R.; Scatena, M.; Gatto, R.; Caruso, S. Molar incisor hypomineralization treatment with casein phosphopeptide and amorphous calcium phosphate in children. *Minerva Stomatol.* **2018**, *67*, 20–25.
7. Atar, M.; Korperich, E.J. Systemic disorders and their influence on the development of dental hard tissues: A literature review. *J. Dent.* **2010**, *38*, 296–306. [CrossRef]
8. Walsh, L.J.; Brostek, A.M. Minimum intervention dentistry principles and objectives. *Aust. Dent. J.* **2013**, *58* (Suppl. 1), 3–16. [CrossRef] [PubMed]
9. Bakry, A.S.; Marghalani, H.Y.; Amin, O.A.; Tagami, J. The effect of a bioglass paste on enamel exposed to erosive challenge. *J. Dent.* **2014**, *42*, 1458–1463. [CrossRef] [PubMed]
10. Bakry, A.S.; Takahashi, H.; Otsuki, M.; Tagami, J. Evaluation of new treatment for incipient enamel demineralization using 45S5 bioglass. *Dent. Mater.* **2014**, *30*, 314–320. [CrossRef] [PubMed]
11. Bakry, A.S.; Takahashi, H.; Otsuki, M.; Sadr, A.; Yamashita, K.; Tagami, J. CO_2 laser improves 45S5 bioglass interaction with dentin. *J. Dent. Res.* **2011**, *90*, 246–250. [CrossRef] [PubMed]
12. Bakry, A.S.; Takahashi, H.; Otsuki, M.; Tagami, J. The durability of phosphoric acid promoted bioglass-dentin interaction layer. *Dent. Mater.* **2013**, *29*, 357–364. [CrossRef] [PubMed]
13. Brauer, D.S.; Karpukhina, N.; O'Donnell, M.D.; Law, R.V.; Hill, R.G. Fluoride-containing bioactive glasses: Effect of glass design and structure on degradation, pH and apatite formation in simulated body fluid. *Acta Biomater.* **2010**, *6*, 3275–3282. [CrossRef] [PubMed]
14. Bakry, A.S.; Al-Hadeethi, Y.; Razvi, M.A. The durability of a hydroxyapatite paste used in decreasing the permeability of hypersensitive dentin. *J. Dent.* **2016**, *51*, 1–7. [CrossRef] [PubMed]
15. Bakry, A.S.; Tamura, Y.; Otsuki, M.; Kasugai, S.; Ohya, K.; Tagami, J. Cytotoxicity of 45S5 bioglass paste used for dentine hypersensitivity treatment. *J. Dent.* **2011**, *39*, 599–603. [CrossRef] [PubMed]
16. Bakry, A.S.; Nakajima, M.; Otsuki, M.; Tagami, J. Effect of Er:YAG laser on dentin bonding durability under simulated pulpal pressure. *J. Adhes. Dent.* **2009**, *11*, 361–368. [PubMed]
17. Bakry, A.S.; Sadr, A.; Inoue, G.; Otsuki, M.; Tagami, J. Effect of Er:YAG laser treatment on the microstructure of the dentin/adhesive interface after acid-base challenge. *J. Adhes. Dent.* **2007**, *9*, 513–520. [PubMed]

18. Bakry, A.S.; Sadr, A.; Takahashi, H.; Otsuki, M.; Tagami, J. Analysis of Er:YAG lased dentin using attenuated total reflectance Fourier transform infrared and X-ray diffraction techniques. *Dent. Mater. J.* **2007**, *26*, 422–428. [CrossRef] [PubMed]
19. Hamba, H.; Nikaido, T.; Sadr, A.; Nakashima, S.; Tagami, J. Enamel lesion parameter correlations between polychromatic micro-CT and TMR. *J. Dent. Res.* **2012**, *91*, 586–591. [CrossRef] [PubMed]
20. Kitasako, Y.; Sadr, A.; Hamba, H.; Ikeda, M.; Tagami, J. Gum containing calcium fluoride reinforces enamel subsurface lesions in situ. *J. Dent. Res.* **2012**, *91*, 370–375. [CrossRef] [PubMed]
21. Bakry, A.S.; Abbassy, M.A. Increasing the efficiency of CPP-ACP to remineralize enamel white spot lesions. *J. Dent.* **2018**, *76*, 52–57. [CrossRef] [PubMed]
22. Hamba, H.; Nikaido, T.; Inoue, G.; Sadr, A.; Tagami, J. Effects of CPP-ACP with sodium fluoride on inhibition of bovine enamel demineralization: A quantitative assessment using micro-computed tomography. *J. Dent.* **2011**, *39*, 405–413. [CrossRef] [PubMed]
23. Reynolds, E.C.; Cai, F.; Cochrane, N.J.; Shen, P.; Walker, G.D.; Morgan, M.V.; Reynolds, C. Fluoride and casein phosphopeptide-amorphous calcium phosphate. *J. Dent. Res.* **2008**, *87*, 344–348. [CrossRef] [PubMed]
24. Tschoppe, P.; Meyer-Lueckel, H. Effects of regular and highly fluoridated toothpastes in combination with saliva substitutes on artificial enamel caries lesions differing in mineral content. *Arch. Oral Biol.* **2012**, *57*, 931–939. [CrossRef] [PubMed]
25. Baysan, A.; Lynch, E.; Ellwood, R.; Davies, R.; Petersson, L.; Borsboom, P. Reversal of primary root caries using dentifrices containing 5000 and 1100 ppm fluoride. *Caries Res.* **2001**, *35*, 41–46. [PubMed]
26. Amaechi, B.T.; Higham, S.M. In vitro remineralisation of eroded enamel lesions by saliva. *J. Dent.* **2001**, *29*, 371–376. [CrossRef]
27. Liu, J.; Rawlinson, S.C.; Hill, R.G.; Fortune, F. Fluoride incorporation in high phosphate containing bioactive glasses and in vitro osteogenic, angiogenic and antibacterial effects. *Dent. Mater.* **2016**, *32*, e221–e237. [CrossRef] [PubMed]
28. Iijima, M.; Onuma, K. Roles of Fluoride on Octacalcium Phosphate and Apatite Formation on Amorphous Calcium Phosphate Substrate. *Cryst. Growth Des.* **2018**, *18*, 2279–2288. [CrossRef]
29. Mneimne, M.; Hill, R.G.; Bushby, A.J.; Brauer, D.S. High phosphate content significantly increases apatite formation of fluoride-containing bioactive glasses. *Acta Biomater.* **2011**, *7*, 1827–1834. [CrossRef] [PubMed]
30. Sugiura, M.; Kitasako, Y.; Sadr, A.; Shimada, Y.; Sumi, Y.; Tagami, J. White spot lesion remineralization by sugar-free chewing gum containing bio-available calcium and fluoride: A double-blind randomized controlled trial. *J. Dent.* **2016**, *54*, 86–91. [CrossRef] [PubMed]

© 2018 by the authors. Licensee MDPI, Basel, Switzerland. This article is an open access article distributed under the terms and conditions of the Creative Commons Attribution (CC BY) license (http://creativecommons.org/licenses/by/4.0/).

Article

Carbonate Apatite Containing Statin Enhances Bone Formation in Healing Incisal Extraction Sockets in Rats

Yunia Dwi Rakhmatia *, Yasunori Ayukawa, Akihiro Furuhashi and Kiyoshi Koyano

Section of Implant and Rehabilitative Dentistry, Division of Oral Rehabilitation, Faculty of Dental Science, Kyushu University, 3-1-1 Maidashi, Higashi-ku, Fukuoka 812-8582, Japan; ayukawa@dent.kyushu-u.ac.jp (Y.A.); furuhasi@dent.kyushu-u.ac.jp (A.F.); koyano@dent.kyushu-u.ac.jp (K.K.)
* Correspondence: rakhmatia@dent.kyushu-u.ac.jp; Tel.: +81-92-642-6441

Received: 14 June 2018; Accepted: 9 July 2018; Published: 12 July 2018

Abstract: The purpose of this study was to evaluate the feasibility of using apatite blocks fabricated by a dissolution–precipitation reaction of preset gypsum, with or without statin, to enhance bone formation during socket healing after tooth extraction. Preset gypsum blocks were immersed in a Na_3PO_4 aqueous solution to make hydroxyapatite (HA) low crystalline and HA containing statin (HAFS), or in a mixed solution of Na_2HPO_4 and $NaHCO_3$ to make carbonate apatite (CO) and CO containing statin (COFS). The right mandibular incisors of four-week-old male Wistar rats were extracted and the sockets were filled with one of the bone substitutes or left untreated as a control (C). The animals were sacrificed at two and four weeks. Areas in the healing socket were evaluated by micro-computed tomography (micro-CT) and histological analyses. The bone volume, trabecular thickness, and trabecular separation were greatest in the COFS group, followed by the CO, HAFS, HA, and C groups. The bone mineral density of the COFS group was greater than that of the other groups when evaluated in the vertical plane. The results of this study suggest that COFS not only allowed, but also promoted, bone healing in the socket. This finding could be applicable for alveolar bone preservation after tooth extraction.

Keywords: carbonate apatite; bone substitute; micro-CT; rat mandibular incisor; tooth extraction

---

1. Introduction

Adequate bone volume and bone density are prerequisites for a predictable long-term prognosis in implant dentistry. Insufficient horizontal or vertical bone in patients precludes the successful outcome of an ideal implant placement [1]. Additional materials, such as autografts, allografts, xenografts, or synthetic bone substitutes are often required to increase and augment the bone volume. In recent years, researchers have developed and fabricated synthetic bone substitutes to achieve a high relative amount of new bone, while avoiding or minimizing the risks of the invasive harvesting of bone from a healthy site, disease transmission, and antigenicity [2].

Calcium sulfate dihydrate ($CaSO_4 \cdot 2H_2O$), known as gypsum, has been approved by the U.S. Food and Drug Administration for clinical use to reconstruct bone defects [3]. Gypsum has the ability to undergo in situ setting after filling the defect, has good biocompatibility, and promotes bone healing [4]. In addition, gypsum can be produced by mixing $CaSO_4 \cdot 0.5H_2O$ powder and water. It is self-setting and can be molded and shaped at room temperature. Gypsum is slightly soluble in water and is thermodynamically unstable in a phosphate-salt-containing solution. It has also been reported that gypsum immersed in a sodium phosphate solution can be transformed to hydroxyapatite [5].

Hydroxyapatite [HA, $Ca_{10}(PO_4)_6(OH)_2$] is considered to be a promising bone substitute in the orthopedic and dental fields because of its high biocompatibility and osteoconductivity [6]. Most HA products are prepared by sintering chemically prepared HA powder at a high temperature. Although the sintering of HA powder provides monolithic HA with good mechanical strength, the crystallinity of the product is too high to be reabsorbed by osteoclasts [7]. To improve this shortcoming, a new method has been proposed to fabricate low-crystalline, porous hydroxyapatite blocks treated with trisodium phosphate solution, using a compositional transformation reaction based on a dissolution–precipitation reaction, with preset gypsum as a precursor [8].

The inorganic component of bone consists of hydroxyapatite with an apatitic crystal solid structure, and contains impurities [9]. The most common impurity is carbonate, which replaces 4–8% of the phosphate groups [10]. In terms of chemical composition, the inorganic component is a carbonated, basic calcium phosphate; hence, it can be termed a carbonate apatite (CO_3Ap: $Ca_{10-a}(PO_4)_{6-b}(CO)_c(OH)_{2-d}$) [11,12]. Sintering is not suitable for the fabrication of CO_3Ap blocks because of the low thermal stability of CO_3Ap at high temperatures, >400 °C [13]. Therefore, a method was proposed to fabricate CO_3Ap blocks by a dissolution–precipitation reaction, with a preset gypsum as an artificially fabricated precursor. Previous studies have described the fabrication on the treatment of preset gypsum with carbonate ion sources added into the system [14,15]. The gypsum blocks were immersed in a mixture of 0.4 mol/L disodium hydrogen phosphate (Na_2HPO_4) and 0.4 mol/L Sodium hydrogen carbonate ($NaHCO_3$) [14]. Sodium hydrogen carbonate and disodium hydrogen phosphate were used as supply sources of CO_3^{2-} and PO_4^{3-} ions [10]. However, another previous study reported that the immersion of preset gypsum in a sodium phosphate solution also produces carbonate apatite, although the carbonate ions are supplied from the atmosphere as CO_2, particularly when the phosphate salt solution is alkaline [16]. The gypsum used as the precursor should have low solubility and must not disintegrate in the solution to allow a balanced dissolution and precipitation process [14]. The fabrication of CO_3Ap blocks in this manner is thought to be a promising artificial bone substitute that mimics bone in terms of chemical inorganic composition.

The mechanism of action of the materials used for bone regeneration is osteoconduction, which provides a scaffold for enhanced bone tissue growth and formation. A promising technique to increase the bioactivity of carbonate apatite blocks is the addition of osteoinductive growth factors or drugs incorporated into the composite. Statins are cholesterol-lowering drugs that inhibit 3-hydroxy-3-methylglutaryl-coenzyme A (HMG-CoA) reductase. A study reported that statin stimulated the bone morphogenetic protein (BMP)-2 expression and showed positive effects on bone formation [17]. Statins have been widely used in alveolar ridge augmentation and bone grafting in the craniofacial region, because of their osteoinductive effect [18–20]. Previous studies reported that the systemic administration of simvastatin promoted bone formation around implants [21] and a topical application of fluvastatin led to bone formation around tibial titanium implants [22]. In addition, the injection of poly(lactic-co-glycolic) acid PLGA-fluvastatin microspheres promoted both bone formation and gingival soft tissue healing [23,24]. Jinno et al. reported that atelo-collagen and alpha-tricalcium phosphate (α-TCP) as a carrier successfully promoted vertical bone formation on the parietal region [25]. Additionally, solutions of statin in optimal concentrations could be combined with bone grafts to enhance their regenerative potential [26,27]. A recent study reported that statin also had antibacterial, antiviral, and antifungal effects that could alter its advantages in clinical dentistry [28].

Dental implant treatment is usually associated with tooth extraction. Bone healing after tooth extraction may prolong the treatment period of 3–6 months. To shorten the treatment period, the preservation of sufficient bone volume and the early healing of alveolar bone following implant placement are desirable. The purpose of the present study was to investigate the effect of statin-containing carbonate apatite and to assess the amount of bone formation induced after the application of this composite in rat incisor extraction sockets.

2. Materials and Methods

2.1. Preparation of Specimens

Commercially available calcium sulfate hemihydrate ($CaSO_4 \cdot 0.5H_2O$, Wako Pure Chemical Industries, Osaka, Japan) was mixed with distilled water at a water to powder ratio of 1:2. For the fluvastatin (FS) group, 0.5 mg FS (Toronto Research Chemicals, North York, Ontario, Canada) was added and mixed with 1 g calcium sulfate hemihydrate paste. The paste was packed into a cylindrical stainless steel mold (6 mm in diameter and 3 mm thick). Both sides of the mold were covered with glass plates and kept at room temperature for 24 h to set the gypsum. The preset gypsum block was then crushed and sieved to obtain 200–400 µm granules.

To make low crystalline apatite, six gypsum granules without FS (HA group) or containing FS (HAFS group) were placed in each vessel (Shikoku Rika, Kochi, Japan) for hydrothermal treatment and immersion in 15 mL of 1 mol/L trisodium phosphate (Na_3PO_4, Wako) aqueous solution, as described previously [8]. The vessels were then placed in an oven (DO.300; As One, Osaka, Japan) at 100 °C for 24 h.

To make the carbonate apatite specimens, the preset gypsum granules without FS (CO group) or containing FS (COFS group) were treated with phosphate and carbonate solution, as described previously [14]. About six gypsum granules from each group were immersed in a 15 mL mixture of 0.4 mol/L disodium hydrogen phosphate (Na_2HPO_4, Wako) and 0.4 mol/L sodium hydrogen carbonate ($NaHCO_3$, Wako), placed in a hydrothermal vessel, and kept at 200 °C for 24 h in a drying oven. After the treatment, the specimens were washed with distilled water and dried at 60 °C for 24 h. The specimen preparation is summarized in Table 1.

Table 1. Summary of preparation of all of the specimens. C—control; HA—hydroxyapatite low crystalline; HAFS—HA containing fluvastatin; CO—carbonate; COFS—CO containing FS.

Sample Groups	$CaSO_4 \cdot 2H_2O$ (Gypsum)	Statin	Immersion Solution	Hydrothermal Treatment
C	X	X	X	X
HA	O	X	Na_3PO_4	100 °C for 24 h
HAFS	O	O	Na_3PO_4	100 °C for 24 h
CO	O	X	Na_2HPO_4 and $NaHCO_3$	200 °C for 24 h
COFS	O	O	Na_2HPO_4 and $NaHCO_3$	200 °C for 24 h

2.2. X-Ray Diffraction Analysis

The specimens were ground to a fine powder and the composition and crystallite size were characterized by X-ray diffraction (XRD) analysis. The XRD patterns were recorded using a powder X-ray diffractometer (D8 Advance A25, Bruker AXS GmbH, Karlsruhe, Germany) with CuKα radiation, operated at a tube voltage of 40 kV and a tube current of 40 mA.

2.3. Scanning Electron Microscope Analysis

The fractured surfaces of the specimens were morphologically evaluated using a scanning electron microscope (SEM; S-3400N, Hitachi High-Technologies, Tokyo, Japan) at an accelerating voltage of 10 kV, after coating with gold-palladium.

2.4. Animals

There were 48 four-week-old male rats that were used in this study; they were fed a commercially-available standard rodent food (CE-2, CLEA Japan, Tokyo, Japan). Water was available ad libitum. The protocol for this study was approved by the Animal Care and Use Committee of Kyushu University (approval number: A-26-064-0).

2.5. Anesthesia and Surgical Procedures

The crown of the mandibular right incisor was cut at the level of the marginal gingiva using a diamond disk with a micromotor handpiece, under anesthesia, every three days prior to extraction so as to loosen the retention by the periodontal ligament and to facilitate the tooth extraction. On the third of the three day periods, the incisor was carefully extracted in a horizontal direction along the long axis of the incisor, under general anesthesia (Figure 1).

In the experimental group, the extracted sockets were filled with 60 mg of either HA, HAFS, CO, or COFS, which was condensed with a root canal plugger using a controlled light force. The sockets were filled to 1 mm short of the orifice in order to avoid infection. In the control (C) group, the sockets were left untreated. At two and four weeks after the incisor extraction and specimen implantation, the animals were deeply anesthetized and perfused with a fixative solution consisting of 0.1 M phosphate-buffered 4% paraformaldehyde (pH 7.4). For a micro-computed tomography (micro-CT) and histological analysis, the right mandibles without soft tissue were dissected out and the samples were fixed in 10% formalin for one week.

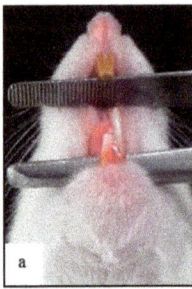
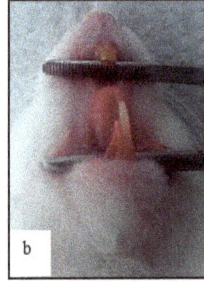
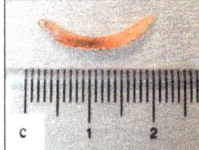

Figure 1. (**a**) Intraoral view after the crown of the mandibular right incisor was cut at the gingival level at 3, 6, and 9 days prior to extraction; (**b**) extraction of the lower right incisor; and (**c**) extracted incisor displaying no signs of fracture.

2.6. Micro-Computed Tomography Analysis

Unprocessed mandibles were imaged and analyzed using an in vivo micro-CT scanner (SkyScan 1076, Aartselaar, Belgium) at 60 kV/167 μA and an Al-0.5 filter. The specimens were fitted into a cylindrical sample holder and scanned in horizontal and vertical positions. High-resolution scanning with a slice thickness of 18 μm was performed. For the micro-CT analysis, a region of interest (ROI) was determined so as to evaluate the socket bone healing in both the horizontal and vertical planes.

The ROI analysis was performed to assess the primary parameters of the bone volume (BV) and the total tissue volume (TV), both measured in mm^3. The TV is the volume of the whole examined sample. This volume is typically defined by a contour or mask, which includes the volume of interest (VOI). The BV was calculated as the volume of the region characterized as bone and normalized ratiometrically against the total volume of the region of interest (BV/TV), in order to derive the percentage bone ratio (%). Bone with different degrees of mineralization (bone mineral density [BMD]) (g/cm^3) records different densities and linear attenuation coefficients, resulting in gray-value variations in the CT scans. Other parameters were trabecular thickness (Tb.Th) to measure the thickness of bone trabeculae (1/mm) and trabecular separation (Tb.Sp) to measure the width of the gap between the bone trabeculae (1/mm).

For the horizontal plane evaluation, the ROI was determined by interpolating the radiographic image on the socket area. For the vertical plane evaluation, the micro-CT scanner software (Version 1.10, Bruker/Skyscan μCT, Kartuizersweg, Kontich, Belgium) was used to make a three-dimensional (3-D) reconstruction from each set of scans. From the entire 3-D data set, an interpolated ROI of the vertical plane was determined, as described previously (Figure 2) [29]. The area of a thickness of 1 mm between the following two planes was observed: the first plane, which was vertical to mandibular plane (plane x), and tangential to the proximal border of the mandibular first molar (plane y), and the second plane, which was parallel and 1 mm medial to the first plane (plane z).

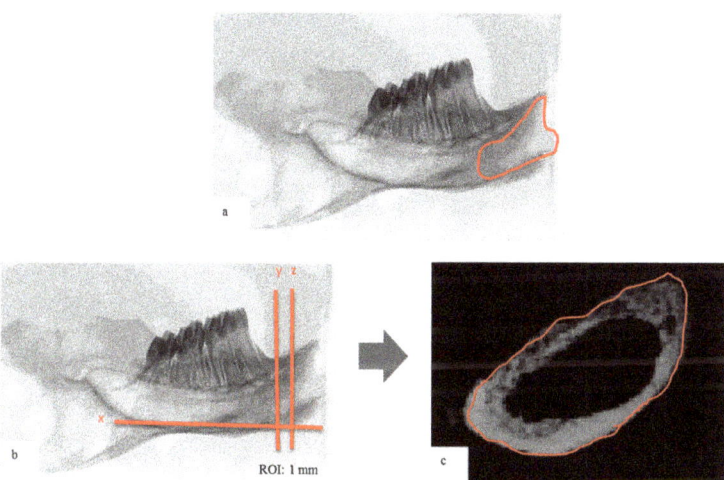

Figure 2. Micro-computed tomography (micro-CT) analysis: (**a**) radiographic image in the horizontal plane; (**b**) radiographic image in the vertical plane (x: mandibular plane, y: plane y; z: plane z), region of interest (ROI) was determined as 1 mm of bone thickness between y and z; (**c**) reconstructed image of ROI before analysis.

2.7. Histological Evaluation

Following the micro-CT scanning, the samples were dehydrated with a graded series of ethanol and were embedded in methacrylate resin. Undecalcified sagittal sections (thickness ~70 µm) were cut, polished, and stained using Masson's trichrome method. For the histological evaluation of the bone and cellular tissue responses, the samples were examined under a light microscope. The center of the test material from one histological section of each specimen was selected to represent that group for evaluation.

2.8. Statistical Analysis

The experimental data were assessed by analysis of variance (ANOVA) with Tukey–Kramer tests for post hoc analysis. The significance level was set at $p < 0.05$. All of the statistical analysis was performed using SPSS 12.0 J (SPSS Japan, Tokyo, Japan).

3. Results

3.1. X-Ray Diffraction Analysis

The XRD patterns of the powdered granules used in this study are summarized in Figure 3. The preset gypsum granules were found to be $CaSO_4 \cdot 2H_2O$ (Figure 3a), and the preset gypsum immersed in Na_3PO_4 solution demonstrated a broad apatitic peak, indicating that it had undergone a compositional transformation from gypsum to low-crystalline apatite (Figure 3b). Similarly, the preset gypsum containing fluvastatin immersed in a Na_3PO_4 solution also demonstrated a broad apatitic peak (Figure 3c).

The preset gypsum with or without fluvastatin and immersed in a solution of Na_2HPO_4 and $NaHCO_3$ (Figure 3d,e) at 200 °C for 48 h also demonstrated a broad apatitic peak. Furthermore, an energy dispersive X-ray spectroscopy analysis of all of the specimens clarified that they did not contain any element except calcium and phosphor, and the X-ray diffraction analysis proved that all of the groups consisted of HA.

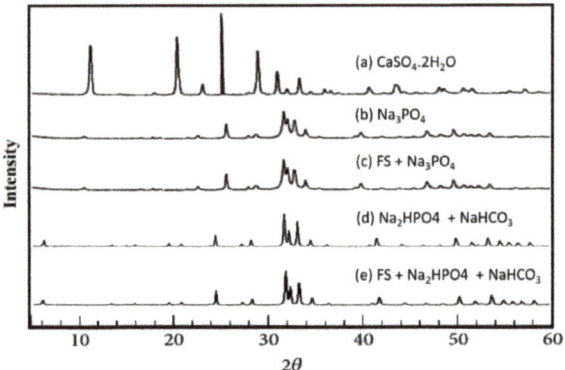

Figure 3. Powder XRD patterns of (**a**) preset gypsum before treatment; (**b**) preset gypsum after immersion in 1 mol/L Na_3PO_4 solution at 100 °C for 24 h; (**c**) preset gypsum containing fluvastatin after immersion in 1 mol/L Na_2HPO_4; (**d**) preset gypsum after immersion in 1 mol/L Na_2HPO_4 and 1 mol/L $NaHCO_3$ at 200 °C for 48 h; and (**e**) preset gypsum containing fluvastatin after immersion in 1 mol/L Na_2HPO_4 and 1 mol/L $NaHCO_3$ at 200 °C for 48 h.

3.2. Scanning Electron Microscope Analysis

SEM images of HA, HAFS, CO, and COFS granules are shown in Figure 4. The density was somewhat higher in the HAFS granules than in the other granules. The HAFS granules retained the morphology of a needle-like gypsum crystal structure covered with many fine granular crystals. The morphology of the CO and COFS granules consisted of tight tangles of needle-like crystals that were smaller and less tangled than those of the HA and HAFS groups.

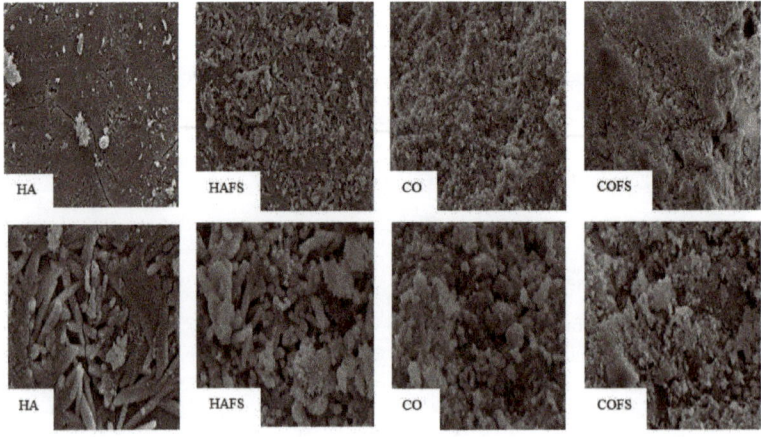

Figure 4. SEM analysis of all groups at magnification ×1000 (upper) and ×4000 (lower).

3.3. Micro-Computed Tomography Analysis

Micro-CT images of all of the groups in the horizontal and vertical planes at two and four weeks after extraction are shown in Figure 5. The micro-CT reconstruction in the vertical plane shows that the most bone formation occurred in the COFS and CO groups, followed by the HAFS and HA groups. More bone formation was observed in the four week groups than the two week groups. Bone growth was observed in the socket area of all of the groups; however, the bone surrounding the socket was thicker in the CO and COFS groups than in the other groups. This result indicates that the carbonate apatite did not hinder the natural bone healing process, but rather enhanced new bone formation in the socket area and the surrounding bone.

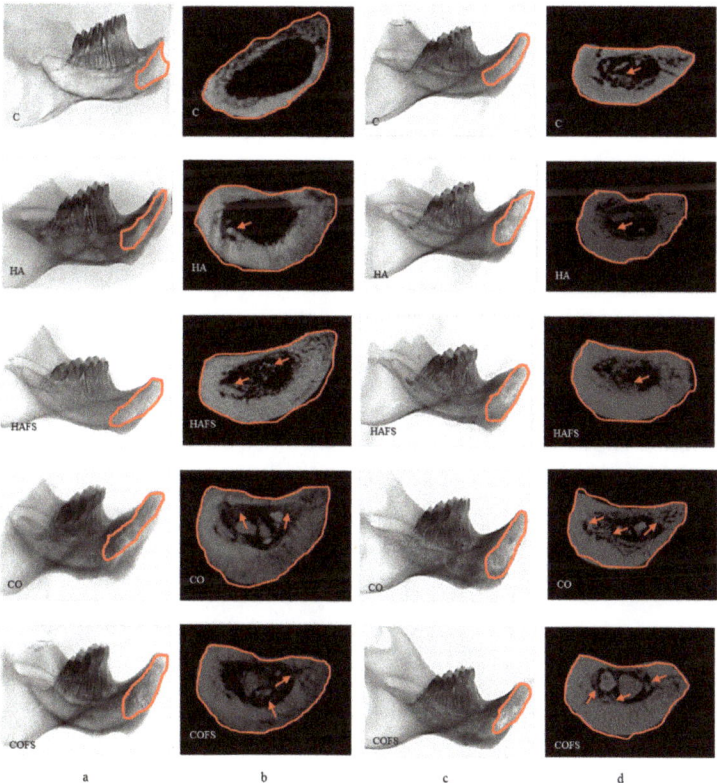

Figure 5. Micro-CT images of the lower right incisor extraction sockets: (**a**) in the horizontal plane at two weeks; (**b**) in the vertical plane at two weeks; (**c**) in the horizontal plane at four weeks; and (**d**) in the vertical plane at four weeks.

The bone volume of the COFS group was greater than that of the other groups, both in the horizontal and vertical planes at two and four weeks after extraction. However, the BMD of the COFS group was lower than that of the HA group in the horizontal plane, but higher in the vertical plane. The difference in the ROI between the horizontal and vertical planes may have caused the differences in the bone volumes of the experimental groups. In addition, the values of Tb.Th and Tb.Sp tended to be higher in the experimental groups than in the control group. Notably, Tb.Th and Tb.Sp were significantly higher in the COFS group than in the CO, HAFS, HA, and control groups at two weeks ($p < 0.05$) (Figure 6).

3.4. Histological Evaluation

New bone formation was observed in all of the sample groups; however, a larger area of bone formation was observed in the COFS group, compared with the other groups, both at two and four weeks (Figure 7). Moreover, a higher mineralization density (as evidenced by the green staining) was observed in the COFS group when compared with the other groups. In the HA group, bone formation was observed in the form of red staining, which defines the lower mineralization density. The histological evaluation was in accordance with the micro-CT results in the vertical plane analysis, although the BMD in the horizontal analysis showed that the COFS group at four weeks tended to have a lower mineralization than the other groups.

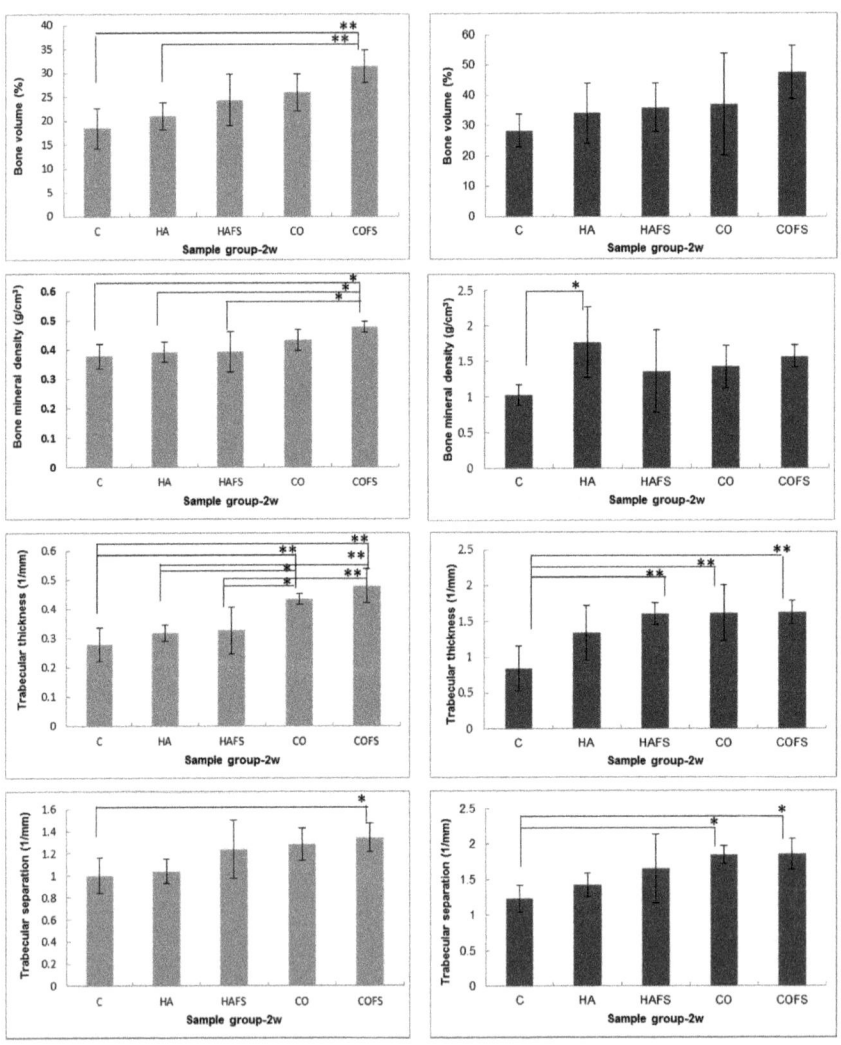

Figure 6. Cont.

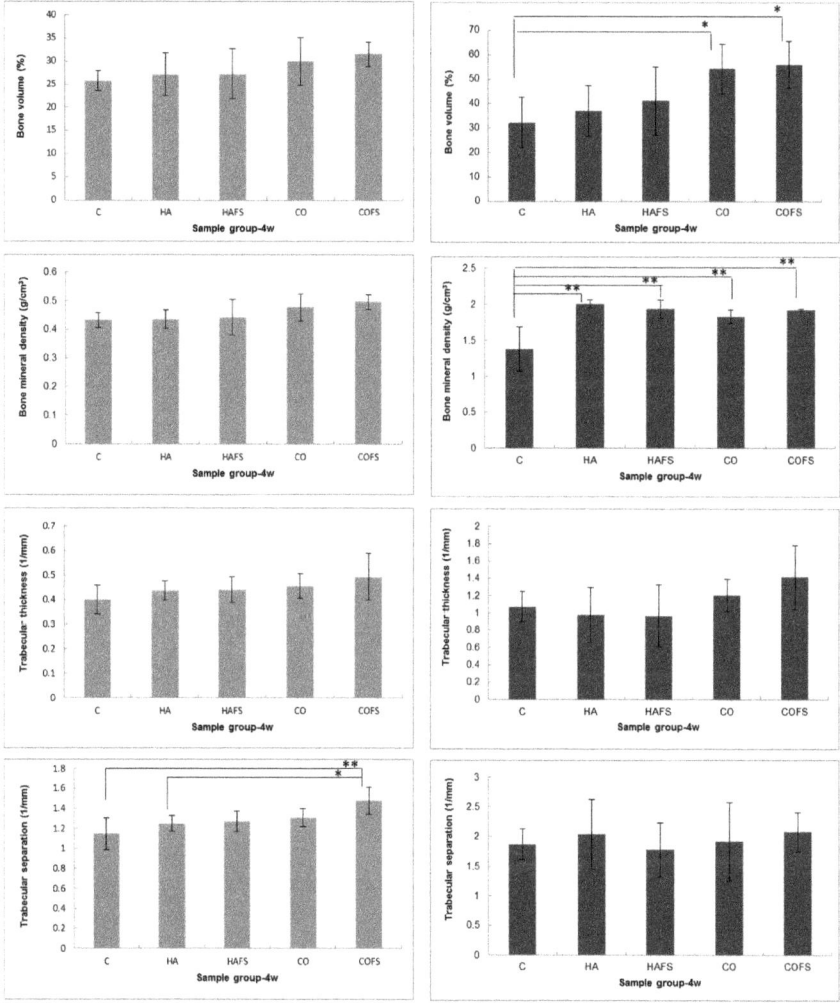

Figure 6. Micro-CT analysis of lower right incisor extraction sockets in vertical (left columns) and horizontal (right columns) planes at two and four weeks: bone volume, bone mineral density, trabecular thickness, and trabecular separation. A p value of <0.05 (*) and a p value of <0.01 (**) were considered significant.

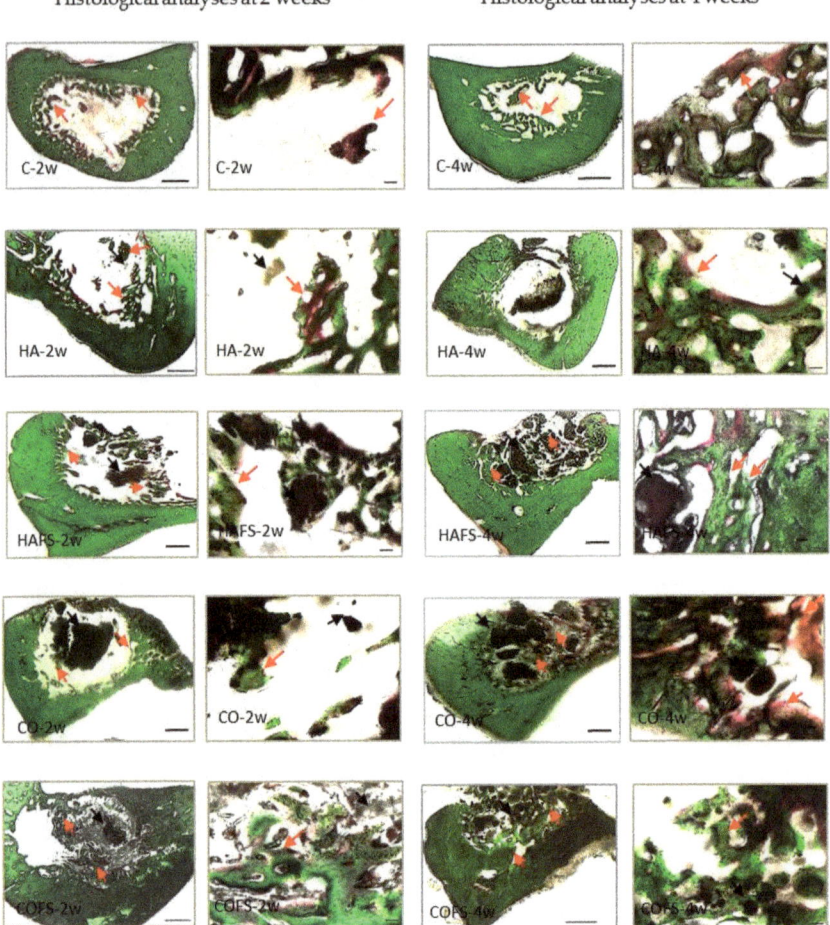

Figure 7. Histological images of lower right incisor extraction sockets at two and four weeks. Red arrows indicate new bone formation and black arrows indicate residual bone substitute in the healing socket. Magnification: ×4 Bar: 500 μm, ×20 Bar: 50 μm.

4. Discussion

The prosthodontic treatment for the edentulous areas, including fixed or removable partial dentures and implants, is strongly related to the extraction socket and residual alveolar bone. The rapid healing potential of the extraction socket has generated interest, because it greatly influences subsequent dental treatment. A human study demonstrated that mineralization begins at the end of the first week, and that most of the granulation tissue has been replaced with a provisional matrix and immature bone by the sixth to eighth week of post-extraction healing [30]. However, because rat extraction sockets are considered to be a non-critical size defect, the socket will be gradually filled by newly-formed bone [29].

Various studies have explored the possibility of accelerating extraction socket healing and preserving the alveolar ridge [29,31]. A commonly used technique is bone grafting, using materials such as bovine bone mineral [32]; β-tricalcium phosphate [33]; or metabolic compounds, such as basic fibroblast growth factor [34] or simvastatin [35]. In this study, the rat sockets were treated with grafted bone as an osteoconductive material and fluvastatin as an osteoinductive material. The materials used to fill the socket area were low crystalline HA and CO_3Ap, without fluvastatin (HA and CO) or with fluvastatin (HAFS and COFS). There were shortcomings of filling these materials in the rat incisor extraction sockets which were approximately 2 mm in diameter and 20 mm in depth; however, the micro-CT results from the experimental groups showed a similar bone formation pattern as the control group (C), indicating that the HA and CO_3Ap, with or without fluvastatin, did not hinder the natural bone healing process in the socket.

The micro-CT analysis was conducted in the horizontal plane to evaluate bone healing in the entire region of the extraction socket and in the vertical plane to establish the median extent of bone healing in the socket area. At two and four weeks, all of the BVs in the horizontal plane were higher than in the vertical plane, which confirmed that bone was formed throughout the whole socket area. The BVs in this study were higher at four weeks than at two weeks after the extraction. Similar previous studies reported that the initial bone formation in rat alveolar wound healing occurred at the end of the first week and that the bone mass increased gradually until the alveolar socket was totally filled with newly formed bone by the 21st day of the healing period [36,37]. However, recent studies reported that bone formation continued to proceed beyond 21 days post extraction, up to the sixth [38] or eighth [39] week. A fundamental step for the subsequent phases of bone healing in the socket area is the existence of blood clot formation [40]. In this study, the materials in the socket may limit the blood clot forming, and so delay the process of bone formation. A similar study reported that bone formation by the second week was delayed in the socket treated with inorganic bone relative to those treated with organic bone. The greater bone volume of the inorganic graft reached a similar amount to that which was observed in the animals grafted with organic bone by the ninth week [41]. The present study has indicated that grafted materials need to be resorbed, and thus, the new bone formation was delayed because of the limitation of blood supply and nutrients in the socket area. Therefore, the marked differences in methodology rendered comparisons between our results and those reported by Okamoto et al. [36] and Vieira et al. [37] unviable.

The BV of the COFS group was significantly higher than that of the other groups at four weeks, followed by the CO and HAFS groups. The results of this study demonstrated that the carbonate content in the COFS group was favored by the inorganic component of the original bone. Additionally, the statin mixed with CO_3Ap and HA was observed to potentially enhance bone formation during socket healing. A previous study reported that there is evidence to suggest that statins, which have been safely used for treatment of hypercholesterolemia, enhance the biosynthesis of BMP-2 [42]. Other studies have reported the positive effects of statins on osteogenesis around implants and in tooth extraction sockets [23,24].

The HA group exhibited a lower level of bone formation. In contrast with this finding, a previous study found a higher level of bone formation with a hydrothermally-treated gypsum soaked in Na_3PO_4 solution than with sintered-HA granules in rat tibia after two weeks [16]. In the reaction to fabricate CO_3Ap, the CO_3^{2-} ions can be supplied in the form of CO_2 from the atmosphere; however, in this study, the CO_3^{2-} was added from a carbonate salt ($NaHCO_3$) to the phosphate salt (Na_2HPO_4) solution. Thus, instead of transforming into CO_3Ap, the hydrothermally-treated gypsum soaked in Na_3PO_4 transformed to low crystalline HA, resulting in a lower level of bone formation. The control group also exhibited minimal bone formation at two weeks, in contrast with a previous study, in which the normal untreated sockets exhibited progressive neo-bone-formation at two weeks [37]. In our preliminary in vitro study, the CO_3Ap released calcium and phosphate ions that induced cell death and affected the osteoblastic activities. Other studies demonstrated that free calcium and inorganic phosphate ions influenced the osteogenic differentiation in vitro of osteoprogenitor cells. Therefore, they suggested making a clear link between the dissolution rate of the calcium phosphate in vitro and early bone formation in vivo [43]. Moreover, the features of the materials in vitro can affect the molecular and cellular interactions at their surface, and consequently can affect the process of bone formation. Differently, the interactions between the implant and its 'biological surrounding' in vivo are highly complex because of the non-equilibrium conditions and because of the undefined amount of compounds playing a role in these interactions [44].

The mineralization in the bone healing process was measured by comparing the BMD of the background bone with HA phantom rods, as part of the micro-CT analysis tools. The degree of mineralization, expressed in milligrams of HA phantom rods per cubic centimeter (mg HA/cm^3), was found to be 0.25 to 0.75 mg HA/cm^3. A threshold of 0.25 mg HA/cm^3 was used to differentiate between the newly formed bone and background bone. Values above 0.75 mg HA/cm^3 were assumed to be graft material. In the vertical plane evaluation at two weeks, the COFS group had a higher BMD than the other groups, but at four weeks, there was no significant difference among all of the groups. In the horizontal plane, the BMD was higher in the HA group than in the control and COFS groups. It is believed that the grafted bone acts as a mineral reservoir inducing bone formation via osteoconductive mechanisms [45]. The mineral content of the HA group was similar to that of the HA phantom rods, resulting in a higher BMD than that of the other groups. In contrast, a previous study using micro-CT reported that a greater bone volume appeared to be linked with lower bone density, probably because the rate of growth of the bone forming cells was greater than the rate of the bone mineralization [46].

The trabecular thickness (Tb.Th) was measured to compare the thickness of the trabecular structures. However, as Tb.Th is a scalar measurement, it may not be able to describe all of the structural changes. The trabecular separation (Tb.Sp) was determined as the mean distance between the mid-axes (i.e., the average separation between the mid-axes). The measurement of Tb.Th and Tb.Sp are related to the diameter of the interconnecting pores. The optimal diameter of the interconnecting pores allows the cells to attach and penetrate through the pores. If the diameter is too small, the cells have difficulty penetrating the structure and will only attach to the outside of the scaffold. In this study, the COFS group had higher Tb.Th and Tb.Sp scores compared with the other groups.

The histological sections of all of the groups supported the quantitative micro-CT findings. Newly-formed trabecular bone was observed mainly on the internal surfaces of the alveolar socket. New bone formation was observed on the surrounding bony walls, but was also seen in the form of 'bony islands' in the places where the socket was filled with grafted bone (HA, HAFS, CO, and COFS), while in the control group, new bone was formed only along the bony walls. Compared with the controls, the treated sockets seemed to contain a smaller amount of blood clot in association with larger amounts of connective tissue and bone. In the four week groups, the treated sockets had a greater relative volume of trabecular bone in parallel with a smaller amount of bone substitute when compared with the two week groups. This finding further supports the biocompatibility and osteoconductivity of HA and CO_3Ap.

Previous studies have shown that the statin in calcium silicate/gypsum/gelatin composite has osteoinductive characteristics, which promote a higher level of bone formation when compared with the grafts without statin [47]. One study reported that the administration of statins increased the expression of bone morphogenetic protein (BMP)-2 mRNA, with a concomitant promotion of bone formation [42]. Statin, with a drug delivery system, has been reported to promote bone formation and soft tissue healing around implants, both through systemic administration or via topical application [21–23]. Statin has also been reported to enhance vascular endothelial growth factor (VEGF) production [48], and to exhibit antimicrobial effects [49]. Thus, statin has the potential to enhance bone formation in volume and quality. In this study, we also demonstrated that CO_3Ap with statin supports a higher level of bone formation, and that it is effective as a bone substitute.

This study did not investigate the ability of the material to be fully absorbed. The bone grafts were observed to remain in the socket area, even after four weeks. However, during the bone remodeling process, osteoclasts produce a weak acidic environment in Howship's lacunae at pH 3–5, to dissolve bone minerals. The solubility of apatite under weak acidic conditions increases with the carbonate content in its apatitic structure [50]. Therefore, CO_3Ap is supposed to be resorbed by osteoclasts.

On this premise, it is concluded that CO_3Ap containing statin is an effective inorganic scaffold for bone regenerative therapy.

5. Conclusions

The present study suggests that gypsum as precursor can be transformed into hydroxyapatite and carbonate apatite. Materials containing statin were proven to be effective to promote bone formation. However, the present study should be regarded with caution because of its limitation of the physical and biological properties of carbonate apatite. Long-term evaluation is required to assess the clinical outcome. Another limitation of the present study was that it was difficult to fulfil the length and depth of the incisor socket area with our materials. Therefore, our data must be interpreted with caution. However, our findings suggest that the osteoconductivity function of CO_3Ap containing fluvastatin promotes bone formation in the early stages of healing in extraction sockets. In light of the new developments in bone regeneration therapeutics, it is important to continue the development of low-cost biomaterials that involve minimal risk during treatment, and that are more reproducible.

Author Contributions: Conceptualization, Y.A. and A.F.; methodology, Y.D.R.; software, Y.A.; validation, Y.D.R., A.F., and Y.A.; formal analysis, Y.D.R.; investigation, A.F.; resources, Y.A.; data curation, Y.D.R.; writing (original draft preparation), Y.D.R.; writing (review and editing), Y.A.; visualization, Y.D.R.; supervision, K.K.; project administration, A.F.; and funding acquisition, Y.D.R.

Funding: This study was supported by JSPS KAKENHI Grant Number JP26861643 from the Japan Society for the Promotion of Science.

Acknowledgments: The authors thank Tya Indah Arifta for her technical assistance in the XRD and SEM analyses. The authors also thank Ikiru Atsuta and Tomohiro Masuzaki for their assistance in histology.

Conflicts of Interest: The authors declare no conflict of interest.

References

1. Rakhmatia, Y.D.; Ayukawa, Y.; Furuhashi, A.; Koyano, K. Current barrier membranes: Titanium mesh and other membranes for guided bone regeneration in dental applications. *J. Prosthodont. Res.* **2013**, *57*, 3–14. [CrossRef] [PubMed]
2. Shetty, V.; Han, T.J. Alloplastic materials in reconstructive periodontal surgery. *Dent. Clin. N. Am.* **1991**, *35*, 521–530. [PubMed]
3. Burg, K.J.L.; Porter, S.; Kellam, J.F. Biomaterial developments for bone tissue engineering. *Biomaterials* **2000**, *21*, 2347–2359. [CrossRef]
4. Favvas, E.P.; Stefanopoulos, K.L.; Vordos, N.C.; Drosos, G.I.; Mitropoulos, A.C. Structural characterization of calcium sulfate bone graft substitute cements. *Mater. Res.* **2016**, *19*, 1108–1113. [CrossRef]
5. Suzuki, Y.; Matsuya, S.; Udoh, K.; Nakagawa, M.; Tsukiyama, Y.; Koyano, K.; Ishikawa, K. Fabrication of hydroxyapatite block from gypsum block based on $(NH_4)_2HPO_4$ treatment. *Dent. Mater. J.* **2005**, *24*, 515–521. [CrossRef] [PubMed]
6. Chang, B.S.; Lee, C.K.; Hong, K.S. Osteoconduction at porous hydroxyapatite with various pore configurations. *Biomaterials* **2000**, *21*, 1291–1298. [CrossRef]
7. Moore, W.R.; Graves, S.E.; Bain, G.I. Synthetic bone graft substitutes. *ANZ J. Surg.* **2003**, *71*, 354–361. [CrossRef]
8. Maruta, M.; Matsuya, S.; Nakamura, S.; Ishikawa, K. Fabrication of low-crystalline carbonate apatite foam bone replacement based on phase transformation of calcite foam. *Dent. Mater. J.* **2011**, *30*, 14–20. [CrossRef] [PubMed]
9. Morgan, E.F.; Barnes, G.L.; Einhorn, T.A. The bone organ system: Form and function. In *Osteoporosis*, 3rd ed.; Marcus, R., Feldman, D., Nelson, D.A., Rosen, C.J., Eds.; Academic Press: San Diego, CA, USA, 2007; pp. 3–26. ISBN 978-0-12-370544-0.
10. Doi, Y.; Aoba, T.; Okazaki, M.; Takahashi, J.; Moriwaki, Y. Analysis of paramagnetic centers in X-ray irradiated enamel, bone, and carbonate-containing hydroxyapatite by electron spin resonance spectroscopy. *Calcif. Tissue Int.* **1979**, *28*, 107–112. [CrossRef] [PubMed]

11. Feng, X. Chemical and biochemical basis of cell-bone matrix interaction in health and disease. *Curr. Chem. Biol.* **2009**, *3*, 189–196. [CrossRef] [PubMed]
12. LeGeros, R.Z. Calcium phosphates in oral biology and medicine. In *Monographs in Oral Science*; Myers, H.M., Ed.; Karger: Basel, Switzerland, 1991; pp. 110–111, ISBN-13: 978-3805552363.
13. Tônsuaadu, K.; Peld, M.; Leskelä, T.; Mannonen, R.; Niinistö, L.; Veiderma, M. A thermoanalytical study of synthetic carbonate containing apatites. *Themochim. Acta* **1995**, *256*, 55–65. [CrossRef]
14. Nomura, S.; Tsuru, K.; Maruta, M.; Matsuya, S.; Takahashi, I.; Ishikawa, K. Fabrication of carbonate apatite blocks from set gypsum based on dissolution-precipitation reaction in phosphate-carbonate mixed solution. *Dent. Mater. J.* **2014**, *33*, 166–172. [CrossRef] [PubMed]
15. Landi, E.; Tampieri, A.; Celotti, G.; Vichi, L.; Sandri, M. Influence of synthesis and sintering parameters on the characteristics of carbonate apatite. *Biomaterials* **2004**, *25*, 1763–1770. [CrossRef] [PubMed]
16. Ayukawa, Y.; Suzuki, Y.; Tsuru, K.; Koyano, K.; Ishikawa, K. Histological comparison in rats between carbonate apatite fabricated from gypsum and sintered hydroxyapatite on bone remodeling. *Biomed. Res. Int.* **2015**, *2015*, 579541. [CrossRef] [PubMed]
17. Montagnani, A.; Gonnelli, S.; Cepollaro, C.; Pacini, S.; Campagna, M.S.; Franci, M.B.; Lucani, B.; Gennari, C. Effect of simvastatin treatment on bone mineral density and bone turnover in hypercholesterolemic postmenopausal women: A 1-year longitudinal study. *Bone* **2003**, *32*, 427–433. [CrossRef]
18. Sonobe, M.; Hattori, K.; Tomita, N.; Yoshikawa, T.; Aoki, H.; Takakura, Y.; Suguro, T. Stimulatory effects of statins on bone marrow-derived mesenchymal stem cells: Study of a new therapeutic agent for fracture. *Bio-Med Mater. Eng.* **2005**, *15*, 261–267.
19. Wong, R.W.K.; Rabie, A.B.M. Early healing pattern of statin induced osteogenesis. *Br. J. Oral Maxillofac. Surg.* **2005**, *43*, 46–50. [CrossRef] [PubMed]
20. Wong, R.W.K.; Rabie, A.B.M. Histologic and ultrastructural study on statin graft in rabbit skulls. *J. Oral Maxillofac. Surg.* **2005**, *63*, 1515–1521. [CrossRef] [PubMed]
21. Ayukawa, Y.; Okamura, A.; Koyano, K. Simvastatin promotes osteogenesis around titanium implants. A histological and histometrical study in rats. *Clin. Oral Implant. Res.* **2004**, *15*, 346–350. [CrossRef] [PubMed]
22. Moriyama, Y.; Ayukawa, Y.; Ogino, Y.; Atsuta, I.; Todo, M.; Takao, Y.; Koyano, K. Local application of fluvastatin improves peri-implant bone quantity and mechanical properities: A rodent study. *Acta Biomater.* **2010**, *6*, 1610–1618. [CrossRef] [PubMed]
23. Masuzaki, T.; Ayukawa, Y.; Moriyama, Y.; Jinno, Y.; Atsuta, I.; Ogino, Y.; Koyano, K. The effect of a single remote injection of statin-impregnated poly(lactic-co-glycolic acid) microspheres on osteogenesis around titanium implants in rat tibia. *Biomaterials* **2010**, *31*, 3327–3334. [CrossRef] [PubMed]
24. Yasunami, N.; Ayukawa, Y.; Furuhashi, A.; Atsuta, I.; Rakhmatia, Y.D.; Moriyama, Y.; Masuzaki, T.; Koyano, K. Acceleration of hard and soft tissue healing in the oral cavity by a single transmucosal injection of fluvastatin-impregnated poly(lactic-co-glycolic acid) microspheres. An in vitro and rodent in vivo study. *Biomed. Mater.* **2015**, *11*, 015001. [CrossRef] [PubMed]
25. Jinno, Y.; Ayukawa, Y.; Ogino, Y.; Atsuta, I.; Tsukiyama, Y.; Koyano, K. Vertical bone augmentation with fluvastatin in an injectable delivery system: A rat study. *Clin. Oral Implant. Res.* **2009**, *20*, 756–760. [CrossRef] [PubMed]
26. Yazawa, H.; Zimmermann, B.; Asami, Y.; Bernimoulin, J.P. Simvastatin promotes cell metabolism, proliferation, and osteoblastic differentiation in human periodontal ligament cells. *J. Periodontol.* **2005**, *76*, 295–302. [CrossRef] [PubMed]
27. Nyan, M.; Sato, D.; Oda, M.; Machida, T.; Kobayashi, H.; Nakamura, T.; Kasugai, S. Bone formation with the combination of simvastatin and calcium sulfate in critical-sized rat calvarial defect. *J. Pharmacol. Sci.* **2007**, *104*, 384–386. [CrossRef] [PubMed]
28. Ting, M.; Whitaker, E.J.; Albandar, J.M. Systematic review of the in vitro effects of statins on oral and perioral microorganisms. *Eur. J. Oral Sci.* **2016**, *124*, 4–10. [CrossRef] [PubMed]
29. Machida, T.; Nyan, M.; Kon, K.; Maruo, K.; Sato, H.; Kasugai, S. Effect of hydroxyapatite fiber material on rat incisor socket healing. *J. Oral Tissue Eng.* **2009**, *7*, 153–162. [CrossRef]
30. Cohen, N.; Cohen-Lévy, J. Healing process following tooth extraction in orthodontic case. *J. Dentofac. Anom. Orthod.* **2014**, *17*, 304. [CrossRef]

31. Kono, T.; Ayukawa, Y.; Moriyama, Y.; Kurata, K.; Takamatsu, H.; Koyano, K. The effect of low-magnitude, high-frequency vibration stimuli on the bone healing of rat incisor extraction socket. *J. Biomech. Eng.* **2012**, *134*, 091001. [CrossRef] [PubMed]
32. Araujo, M.G.; Liljenberg, B.; Lindhe, J. Dynamics of bio-oss collagen incorporation in fresh extraction wounds: An experimental study in the dog. *Clin. Oral Implant. Res.* **2010**, *21*, 55–64. [CrossRef] [PubMed]
33. Araujo, M.G.; Liljenberg, B.; Lindhe, J. Beta-tricalcium phosphate in the early phase of socket healing: An experimental study in the dog. *Clin. Oral Implant. Res.* **2010**, *21*, 445–454. [CrossRef] [PubMed]
34. Fei, W.; Zuo, L.; Hu, J.; Yin, M.; Shen, Z. Basic fibroblast growth factor accelerates wound healing of tooth extraction. *Chin. J. Oral Implantol.* **2001**, *4*, 154–156.
35. Wu, Z.; Liu, C.; Zang, G.; Sun, H. The effect of simvastatin on remodelling of the alveolar bone following tooth extraction. *Int. J. Oral Maxillofac. Surg.* **2008**, *37*, 170–176. [CrossRef] [PubMed]
36. Okamoto, T.; Onofre Da Silva, A. Histological study on the healing of rat dental sockets after partial removal of the buccal bony plate. *J. Nihon Univ. Sch. Dent.* **1983**, *25*, 202–213. [CrossRef] [PubMed]
37. Vieira, A.E.; Repeke, C.E.; Ferreira Junior Sde, B.; Colavite, P.M.; Biguetti, C.C.; Oliveira, R.C.; Assis, G.F.; Taga, R.; Trombone, A.P.; Garlet, G. Intramembranous bone healing process subsequent to tooth extraction in mice: Micro-computed tomography, histomorphometric and molecular characterization. *PLoS ONE* **2015**, *10*, e0128021. [CrossRef] [PubMed]
38. Lamano Carvalho, T.L.; Bombonato, K.F.; Brentegani, L.G. Histometric analysis of rat alveolar wound healing. *Braz. Dent. J.* **1997**, *8*, 9–12.
39. Elsubeihi, E.S.; Heersche, J.N. Quantitative assessment of post-extraction healing and alveolar ridge remodeling of the mandible in female rats. *Arch. Oral Biol.* **2004**, *49*, 401–412. [CrossRef] [PubMed]
40. Yugoshi, L.I.; Sala, M.A.; Brentegani, L.G.; Lamano Carvalho, T.L. Histometric study of socket healing after tooth extraction in rats treated with diclofenac. *Braz. Dent. J.* **2002**, *13*, 92–96. [CrossRef] [PubMed]
41. Calixto, R.F.; Teófilo, J.M.; Brentegani, L.G.; Lamano, T. Comparison of rat bone healing following intra-alveolar grafting with organic or inorganic bovine bone particles. *Braz. J. Oral Sci.* **2008**, *7*, 1512–1519.
42. Mundy, G.; Garrett, R.; Harris, S.; Chan, J.; Chen, D.; Rossini, G.; Boyce, B.; Zhao, M.; Gutierrez, G. Stimulation of bone formation in vitro and rodents by statins. *Science* **1999**, *286*, 1946–1949. [CrossRef] [PubMed]
43. De Bruijn, J.D.; Bovell, Y.P.; van Blitterswijk, C.A. Structural arrangements at the interface between plasma sprayed calcium phosphates and bone. *Biomaterials* **1994**, *15*, 543–550. [CrossRef]
44. Dhert, W.J.; Thomsen, P.; Blomgren, A.K.; Esposito, M.; Ericson, L.E.; Verbout, A.J. Integration of press-fit implants in cortical bone: A study on interface kinetics. *J. Biomed. Mater. Res.* **1998**, *41*, 574–583. [CrossRef]
45. Wagner, J.R. Clinical and histological case study using resorbable hydroxylapatite for the repair of osseous defects prior to endosseous implant surgery. *J. Oral Implantol.* **1989**, *15*, 186–192. [PubMed]
46. Rakhmatia, Y.D.; Ayukawa, Y.; Jinno, Y.; Furuhashi, A.; Koyano, K. Micro-computed tomography analysis of early stage bone healing using micro-porous titanium mesh for guided bone regeneration: Preliminary experiment in a canine model. *Odontology* **2017**, *105*, 408–417. [CrossRef] [PubMed]
47. Zhang, J.; Wang, H.; Shi, J.; Wang, Y.; Lai, K.; Yang, X.; Chen, X.; Yang, G. Combination of simvastatin, calcium silicate/gypsum, and gelatin and bone regeneration in rabbit calvarial defects. *Sci. Rep.* **2016**, *6*, 23422. [CrossRef] [PubMed]
48. Bitto, A.; Minutoli, L.; Altavilla, D.; Polito, F.; Fiumara, T.; Marini, H.; Galeano, M.; Calò, M.; Lo Cascio, P.; Bonaiuto, M.; et al. Simvastatin enhances VEGF production and ameliorates impaired wound healing in experimental diabetes. *Pharmacol. Res.* **2008**, *57*, 159–169. [CrossRef] [PubMed]
49. Jerwood, S.; Cohen, J. Unexpected antimicrobial effect of statins. *J. Antimicrob. Chemother.* **2008**, *61*, 362–364. [CrossRef] [PubMed]
50. LeGeros, R.Z.; Ming, S. Chemical stability of carbonate- and fluoride-containing apatites. *Caries Res.* **1983**, *17*, 419–429. [CrossRef] [PubMed]

© 2018 by the authors. Licensee MDPI, Basel, Switzerland. This article is an open access article distributed under the terms and conditions of the Creative Commons Attribution (CC BY) license (http://creativecommons.org/licenses/by/4.0/).

Article

Osteoblast Cell Response to Naturally Derived Calcium Phosphate-Based Materials

Valentina Mitran [1], Raluca Ion [1], Florin Miculescu [2,*], Madalina Georgiana Necula [1], Aura-Catalina Mocanu [2,3], George E. Stan [4], Iulian Vasile Antoniac [2] and Anisoara Cimpean [1,*]

1 Department of Biochemistry and Molecular Biology, University of Bucharest, 91-95 Spl. Independentei, 050095 Bucharest, Romania; valentinamitran@yahoo.com (V.M.); rciubar@yahoo.com (R.I.); necula.madalina92@gmail.com (M.G.N.)
2 Department of Metallic Materials Science, Physical Metallurgy, University Politehnica of Bucharest, 313 Splaiul Independentei, J Building, District 6, 060042 Bucharest, Romania; mcn_aura@hotmail.com (A.-C.M.); antoniac.iulian@gmail.com (I.V.A.)
3 S.C. Nuclear NDT Research & Services S.R.L, Department of Research, Development and Innovation, 104 Berceni Str., Central Laboratory Building, District 4, 041912 Bucharest, Romania
4 National Institute of Materials Physics, Laboratory of Multifunctional Materials and Structures, Atomistilor Str., No. 405A P.O. Box MG 7, 077125 Măgurele-Bucharest, Romania; george_stan1@yahoo.com
* Correspondence: f_miculescu@yahoo.com (F.M.); anisoara.cimpean@bio.unibuc.ro (A.C.); Tel.: +40-21-3169563 (F.M.); +40-21-3181575 (ext. 106) (A.C.)

Received: 6 June 2018; Accepted: 25 June 2018; Published: 27 June 2018

Abstract: The demand of calcium phosphate bioceramics for biomedical applications is constantly increasing. Efficient and cost-effective production can be achieved using naturally derived materials. In this work, calcium phosphate powders, obtained from dolomitic marble and *Mytilus galloprovincialis* seashells by a previously reported and improved Rathje method were used to fabricate microporous pellets through cold isostatic pressing followed by sintering at 1200 °C. The interaction of the developed materials with MC3T3-E1 pre-osteoblasts was explored in terms of cell adhesion, morphology, viability, proliferation, and differentiation to evaluate their potential for bone regeneration. Results showed appropriate cell adhesion and high viability without distinguishable differences in the morphological features. Likewise, the pre-osteoblast proliferation overtime on both naturally derived calcium phosphate materials showed a statistically significant increase comparable to that of commercial hydroxyapatite, used as reference material. Furthermore, evaluation of the intracellular alkaline phosphatase activity and collagen synthesis and deposition, used as markers of the osteogenic ability of these bioceramics, revealed that all samples promoted pre-osteoblast differentiation. However, a seashell-derived ceramic demonstrated a higher efficacy in inducing cell differentiation, almost equivalent to that of the commercial hydroxyapatite. Therefore, data obtained demonstrate that this naturally sourced calcium-phosphate material holds promise for applications in bone tissue regeneration.

Keywords: osteoblast; biocompatibility; naturally derived calcium phosphate; seashell; dolomitic marble

1. Introduction

The understanding of processes involved in bone biomineralization has led lately to the development of improved biomimetic synthesis methods and the production of a new generation of biomaterials. Calcium carbonate ($CaCO_3$) emerged as a sustainable bioresource for various calcium phosphates synthesis along with coralline hydroxyapatite [1] and soon became a target subject for the extensive research in the reconstructive orthopedics field. Now both the marine (corals, seashells, cockleshells, cuttlefish bones) and terrestrial forms (marble, land snails) of $CaCO_3$ are known as an eco-friendly, sustainable and geographically available resource, but improvements are still necessary in terms of calcium phosphates synthesis parameters [2]. The use of natural materials and structures for medical purposes was motivated by the limitations of synthetic materials generation in regard to the necessary mechanical features and integrity [3–6]. In terms of targeted precursors, dolomitic marble and *Mytilus galloprovincialis* seashells, autochthonously available, represent two different calcium carbonate polymorphic forms—calcite and aragonite—which have similar chemical composition, except for the Mg ion content that defines the dolomitic marble species [2]. Research carried out has shown that both marine and terrestrial resources can be an appropriate resource for biological phase-pure and thermally stable CPC (calcium phosphate ceramics) production [7].

Given that the mineral component of the human bone is apatite, calcium phosphate materials have been readily preferred because of their role in the bone remodeling process. In terms of the chemical composition, calcium phosphates currently used as biomaterials are classified as: calcium hydroxyapatite (HA), α and β tricalcium phosphate (α- and β-TCP), biphasic calcium phosphates, represented by mixtures of HA and β-TCP, and unsintered or calcium-deficient apatite [8].

The use of CPC provides advantages in terms of new bone formation processes, which are influenced by crystallinity and the Ca/P mole ratio, strongly related to the release of calcium and phosphate ions required for bone mineralization. In addition, extensive literature has shown that calcium phosphate bioceramic pellets promote both osteogenesis and osteointegration, in correlation to the samples' chemistry and surface load, as well as topography [9–11]. However, it should be noted that the target application of these ceramics is to transiently replace natural bone [9].

The growth of osteoblasts on such surfaces requires certain dimensions and an interconnected pore structure. Complementary to the physical structure, chemical composition affects the performance of HA ceramics, so that large crystal decks and high CaO content decrease biocompatibility [12]. At the same time, HA obtained from natural sources is non-stoichiometric and may incorporate other ions, for example CO_3^{2-}, traces of Fe^{2+}, Na^+, Mg^{2+}, F^-, and Cl^-, so making them more similar to natural apatite compared to pure stoichiometric HA, and therefore more bioactive [13]. Hence, the presence of these ions influences numerous biochemical reactions related to bone metabolism. For instance, the presence of Na^+ and Mg^{2+} is very important for bone development, their absence leading to bone loss and fragility [14]. Possible disadvantages of these materials relate to the differences regarding Ca/P mole ratio, particle size, morphology, phase composition, thermal stability or trace elements composition when different sources are used or due to batch variation [15].

HA is a calcium phosphate ceramic with a similar crystallographic structure to the naturally sourced apatite and has proved successful when involved in surgical and dental bone defects restoration for the last 20 years [16,17]. Besides, HA exhibits excellent biological and chemical affinity for bone tissue, and animal studies have shown that this ceramic is in the long-term biocompatible without harmful side effects, immunogenicity or inflammatory responses [18–24]. From a thermodynamic point of view, HA is the most stable calcium phosphate ceramic [25]. However, the costs associated with the synthesis of HA from inorganic Ca- and P-based sources are frequently high. Hence, an increasing number of articles have been devoted to the synthesis, characterization, and application of naturally derived HA, which has demonstrated superior biological behavior when compared to chemically synthesized HA.

In this context, the aim of this study was to evaluate the biological properties of calcium phosphate-based materials derived from dolomitic marble and *Mytilus galloprovincialis* seashell in comparison with those of commercial hydroxyapatite. The in vitro response of MC3T3-E1 pre-osteoblasts to the developed naturally derived materials revealed their ability to support cell adhesion and viability, maturation of the actin cytoskeleton, as well as cell proliferation and differentiation.

2. Materials and Methods

2.1. Ceramic Synthesis

The tested samples for this research were synthesized through a parametrically improved Rathje method [26]. The experimental setup involved the following stages: (i) thermal dissociation of autochthonous seashells (*Mytilus galloprovincialis*, $CaCO_3$) and dolomitic marble ($CaMg(CO_3)_2$) resulting in calcium oxide (CaO) powder was carried out; (ii) after hydration of the CaO, the resultant calcium hydroxide ($Ca(OH)_2$) powder (10 g) was mixed with 200 mL distilled water and a stoichiometrically calculated amount of phosphoric acid (5.5 mL), was added dropwise at room temperature; (iii) magnetic stirring of the obtained slurry was conducted for 2 h and (iv) aging and drying of the powder was carried out at room temperature and in an electric oven (200 °C), respectively [3,27]. The resultant ceramic powders were ground in a planetary mill with an agate bowl and balls and sorted using standardized granulometric sieves (<100 µm particle size). They were further subjected to cold isostatic pressing with a force of 10 MPa in a Φ30 mm mold. Pellets were then kept at room temperature for 24 h and sintered at 1200 °C for 10 h. Subsequently, samples were ground using abrasive paper P2500 to obtain parallel planar surfaces. Some of them were exposed to the three-point bending test as a method for fractographic surfaces achievement.

2.2. Characterization of the Synthesized Products

For both the synthesis and sintering stages, products were investigated through:

- Morpho compositional analysis: by Scanning Electron Microscopy—Philips Xl 30 ESEM TMP (FEI/Phillips, Hillsboro, OR, USA) coupled with Energy Dispersive Spectroscopy—EDAX Sapphire spectrometer. An acceleration voltage of 25 kV and working distance of 10 mm were used. The SEM and EDS investigations were conducted on three randomly chosen areas.
- Structural analyses: by (1) X ray diffraction, XRD—Bruker D8 (Bruker Corporation, Billerica, MA, USA). An Advance diffractometer was used equipped with a LynxEye detector, in Bragg-Brentano geometry, with Cu K_α (λ = 1.5418 Å) radiation with the scattered intensity being scanned in the 2θ range of 10–60°, with a step size of 0.04° and a dwell time of 1 s; and (2) Fourier Transform Infrared Spectroscopy: FT-IR were acquired on a Perkin Elmer Spectrum BX II spectrometer (PerkinElmer, Inc., Waltham, MA, USA), in attenuated total reflectance (ATR) mode (PikeMiracle head); the spectra being recorded over the range 4000–500 cm^{-1}, at a resolution of 4 cm^{-1} and using 32 scans/experiment.

As a comparison measure, we chose a commercial hydroxyapatite (Merck KGaA, Darmstadt, Germany) as a reference material. Preliminary results had already been published [3,27–30], but were confirmed by the new investigations carried out in the present study.

After thermal treatment, wettability was evaluated by contact angle (CA) measurements with 3 different wetting agents: water, diiodomethane (DIM) and ethylene glycol (EG), using a Krüss Drop Shape Analyser—DSA100 (A. Krüss Optronic GmbH, Hamburg, Germany). The experiments were conducted at constant parameters of 20 ± 1 °C temperature and room humidity of 45 ± 5% with images being captured at 1 s after wetting agent droplet deposition. The results comprised an average of 3 determinations/sample.

2.3. Cell Culture Experiments

Mouse pre-osteoblasts (MC3T3-E1, ATCC®, CRL-2593™) were directly seeded on the top of the ceramic samples (marble- and seashell-derived calcium phosphate-based materials, and commercial HA) at a density of 1×10^4 cells·cm^{-2} excepting the osteogenic differentiation studies where an initial cell density of 4×10^4 cells·cm^{-2} was used. Prior to cell seeding, the tested samples (smaller discs with a diameter of 10 mm and an area of 0.785 cm^2) were sterilized at 180 °C for 1 h, washed four times for 15 min with Dulbecco's Modified Eagle's Medium (DMEM, Sigma-Aldrich Co., St. Louis, MO, USA) and then placed in 12-well plates. After that, the cells attached to the substrates were incubated in DMEM supplemented with 10% fetal bovine serum (Gibco (Life Technologies Corporation, Grand Island, NY, USA)) and 1% (v/v) penicillin/streptomycin (10,000 units mL^{-1} penicillin and 10 mg mL^{-1} streptomycin) (Sigma Aldrich) in a humidified atmosphere of 5% CO_2 at 37 °C for specific time points. Pre-osteoblast differentiation was assayed both in standard culture medium and osteogenic medium containing ascorbic acid (AA, 50 µg/mL) and β-glycerophosphate (β-GP, 5 mM). All experiments were performed in triplicate.

2.3.1. Assessment of the Cellular Survival and Proliferation

Possible cytotoxic effects of the analyzed samples were evaluated by quantification of lactate dehydrogenase (LDH) released in the culture medium by the cells grown in contact with their surfaces. This test was performed after 1 and 4 days of culture, by using a "LDH-based In Vitro Toxicology Assay Kit" (Sigma-Aldrich Co. St. Louis, MO, USA), according to the manufacturer's protocol. Absorbance was recorded at 490 nm using a microplate reader (Thermo Scientific Appliskan, Vantaa, Finland), low OD values indicating the materials' capacity to sustain cellular survival.

Cell viability/ proliferation was quantified at the same time points by means of MTT (3-(4,5-Dimethyl-2-thiazolyl)-2,5-diphenyl-2H-tetrazolium bromide) assay. At specified time points the cells were incubated with 1 mg mL^{-1} MTT solution for 3 h at 37 °C. Then, the MTT solution was removed and the insoluble purple formazan produced by metabolically active viable cells was solubilized with dimethyl sulfoxide and the absorbance of dye measured at 550 nm using a microplate reader (Thermo Scientic Appliskan, Vantaa, Finland).

The above quantitative assays were accompanied by a qualitative assay consisting of cell staining with a LIVE/DEAD Cell Viability/Cytotoxicity Assay Kit (Molecular Probes, Eugene, OR, USA). This assay was performed in accordance with the manufacturer's instructions. Briefly, after 1 and 4 days of culture, the analyzed samples were incubated in a solution containing 2 mM calcein-AM and 4 mM ethidium homodimer-1 (EthD-1) for 10 min at room temperature, in the dark. Afterwards, the samples were washed with phosphate buffered saline (PBS, Gibco) and examined under an inverted microscope Olympus IX71 (Olympus, Tokyo, Japan) to detect viable (green fluorescence) and dead (red fluorescence) cells. The fluorescent images were captured using a Cell F Image acquiring system.

2.3.2. Evaluation of MC3T3-E1 Cell Adhesion and Morphology

The adhesion and morphological appearance of MC3T3-E1 cells grown on each ceramic sample were assessed by immunoreactive staining of vinculin and labeling of actin filaments with phalloidin coupled with Alexa Fluor 546, at 3 h and 24 h post-seeding. In this assay, the pre-osteoblasts grown on the substrates were fixed with 4% paraformaldehyde, permeabilized and blocked with 0.1% Triton X-100/2% Bovine Serum Albumin (BSA) in PBS and incubated with mouse anti-vinculin monoclonal antibody, dilution 1:50 (Santa Cruz Biotechnology, Dallas, TX, USA) in PBS containing 1.2% BSA. After washes with PBS, they were further incubated with Alexa Fluor 488-conjugated goat anti-mouse IgG antibody (Invitrogen, Eugene, OR, USA) in PBS containing 1.2% BSA for 1 h. After this step, phalloidin conjugated with Alexa Fluor 546, 20 µg/mL (Invitrogen, Eugene, OR, USA) was added for actin staining. Also, the nuclei were marked with 2 µg/mL 6-diamidino-2-phenylindole (DAPI, Sigma-Aldrich Co., Steinheim, Germany). Labeled samples were washed with PBS and examined under an inverted microscope equipped with epifluorescence (Olympus IX71, Olympus, Tokyo, Japan). Representative, microscopic images were captured using the Cell F software.

2.3.3. Measurement of the Pre-Osteoblast Differentiation

To determine whether the three synthesized ceramics supported differentiation of MC3T3-E1 cells in osteoblasts, ALP activity and collagen synthesis were measured under both standard and osteogenic culture conditions. The intracellular ALP activity was determined after 7 and 14 days of culture by using Alkaline Phosphatase Activity Colorimetric Assay Kit (BioVision, Milpitas, CA, USA), as we reported in a recent paper [31]. Briefly, 80 µL from cellular lysate were mixed with 50 µL of 5 mM p-nitrophenylphosphate (pNPP) and incubated for 1 h at room temperature in the dark. After this step, 20 μL of stop solution was added to each sample and the absorbance was measured at 405 nm using a microplate reader (Thermo Scientific Appliskan, Vantaa, Finland). A standard curve was used to determine the concentrations of the reaction product. ALP activity was calculated using the formula: ALP activity (U/mL) = A/V/T, where A represents the amount of p-nitrophenol (pNP) expressed by the samples (in mol), V is the volume of cell lysate used in reaction (in mL) and T is the reaction time (in min). Also, the protein concentrations were measured for each sample using the Bradford reaction and ALP activity was normalized to 1 μg of protein.

The measurement of collagen synthesis and deposition on the cell supporting biomaterials was performed at 3-weeks post-seeding by staining with Sirius Red, as previously described [32]. Briefly, samples were washed with PBS and fixed with 10% paraformaldehyde. After three washes with deionized water, samples were maintained in 0.1% solution of Sirius Red (Bio-Optica, Milano, Italy) for 1 h at room temperature and washed again. Next, samples were dried for 24 h in air. Finally, the stain was dissolved in 0.2 M NaOH/methanol (1:1) and the optical density was recorded at 540 nm.

2.4. Statistical Analysis

Statistical analysis of data was performed with GraphPad Prism software (Version 3.03, GraphPad, San Diego, CA, USA) using one-way ANOVA with Bonferroni's multiple comparison tests. All values are expressed as means ± standard deviation (SD) of three independent experiments and differences at $p < 0.05$ were considered statistically significant.

3. Results and Discussion

3.1. Materials' Characterization

After thermal dissociation and hydration stages, chemical analysis proved the presence of only calcium hydroxide characteristic elements and the conservation of Mg composition for marble-derived samples. Microstructure analysis of calcium hydroxide powder (Figure 1a,b) revealed the tendency of fine particles to agglomerate as irregularly sized and shaped crystals (1–10 μm). Compared to the regular and fine microstructure of Merck HA, marble and seashell derived bioceramics (Figure 1c–e) displayed mainly clustered grains and few dispersed polyhedral and needle-like ones with variable sizes in the range of 1–20 μm. The sintering process induced an irregular micrometric layered morphology with a high compaction degree and accentuated porosity (~1 μm min. pore diameter), similar to the reference samples. Supplementary, fractography analysis indicated a brittle intercrystalline fracture of both sample types (Figure 1f–h) with prominent and defined grain boundaries. In addition, it revealed a porous structure, with augmented internal microporosity (0.5–1 μm pore diameter), far more accentuated for seashell derived samples. After thermal treatment pellet shrinkage of linear dimensions was estimated at ~15% (samples diameter of ~26.5 mm and thickness of 3 mm).

Post synthesis, EDS results (see Table 1) exposed a typical calcium phosphate composition and values ranging from 1.50–1.53 for the Ca/P mole ratio. This correlated with the XRD patterns revealing a biphasic composition (HA and brushite) only for marble-derived samples, which was expected given the inhibitory effect of Mg on HA precipitation. More to the point, FT-IR spectra confirmed these results as for all samples, HA characteristic bands were detected and in particular, more intense peaks corresponding to hydrogen phosphate ions were attributed to brushite presence. Further thermal treatment improved and modulated the chemical and structural composition. As a key indicator for derived bioceramics performance, the Ca/P mole ratio increased from 1.60–1.62, which is close to the reference value of HA (1.67), with no alteration of the elemental composition. In terms of structure, given the high temperature treatment, both XRD and FT-IR analyses (Figures 2 and 3) showed a biphasic transformation for all synthesized materials: i.e. with HA coexisting with β-TCP and with no trace of residual, cytotoxic compounds [3].

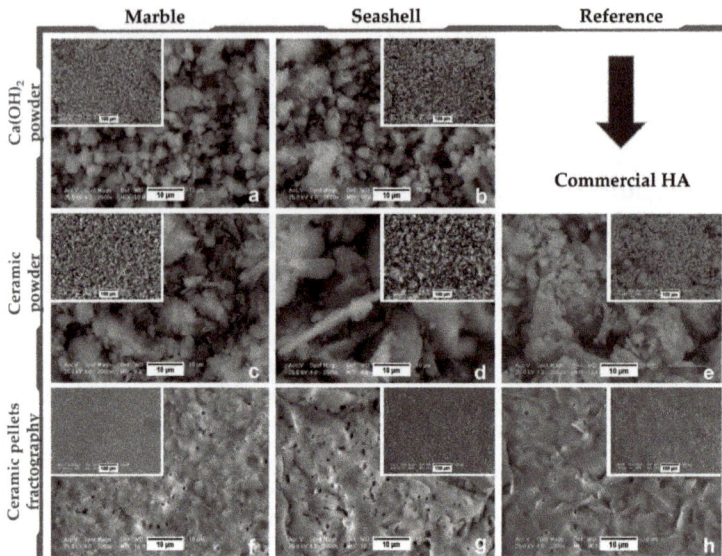

Figure 1. Representative micrographs for marble, seashell and reference material derived products: Ca(OH)$_2$ (**a,b**), Ceramic powder (**c–e**), Ceramic pellets fractographic surface (**f–h**).

Table 1. Chemical composition of naturally derived calcium phosphate based-materials: (A) post Synthesis and (B) post Thermal Treatment and compressive strength of sintered pellets.

Sample Type	O (at. %)		Mg (at. %)		P (at. %)		Ca (at. %)		Ca/P ratio		Compressive Strength (N/mm^2)
	(A)	(B)	(A)	(B)	(A)	(B)	(A)	(B)	(A)	(B)	
Marble	55.10	54.04	0.58	1.28	17.71	16.65	26.61	26.64	1.50	1.60	2.37
Seashell	60.76	55.99	-	-	15.51	15.81	23.73	25.61	1.53	1.62	4.53

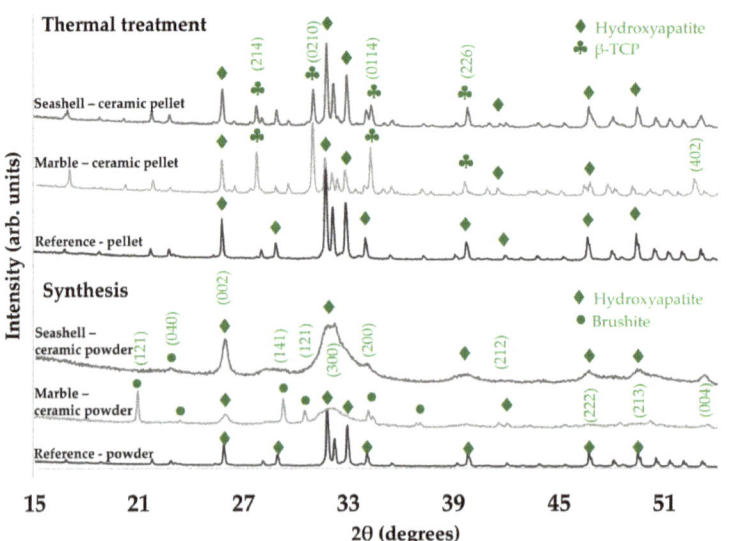

Figure 2. Representative XRD patterns for seashell, marble and reference material derived products.

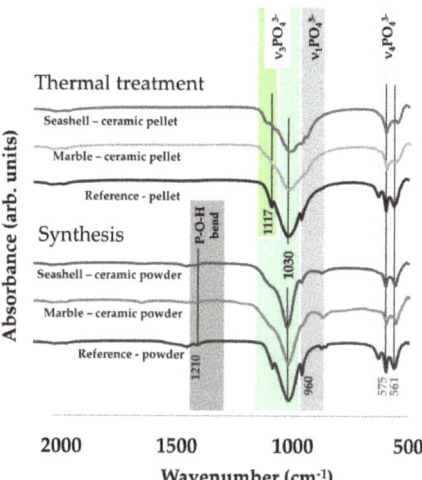

Figure 3. Representative FT-IR spectra for seashell, marble and reference material derived products.

Taken together, the results summarize the potential of the adapted Rathje method for naturally derived calcium phosphates synthesis, even from marble precursors, and the importance of thermal treatment for modulated, enhanced morpho-compositional, and structural features.

Regarding the wettability of the sintered pellets, contact angle (CA) values varied from 40–51° and 45–61° for marble and seashell derived samples, respectively, and 41–57° for reference material (Figure 4). Therefore, results revealed a hydrophilic character (CA values < 90°) independent of the wetting agent and natural precursor, similar to the reference sample. However, given that surface morphology is a key factor for CA [33], the lowest values correspond to samples with an accentuated surface porosity (see Figure 1) and better surface wetting properties. Moreover, the ascending or descending trendline of CA found for marble and seashell derived samples is strongly related to the microporosity range of the samples, the surface preferential behavior to the wetting agent and the samples' anisotropy.

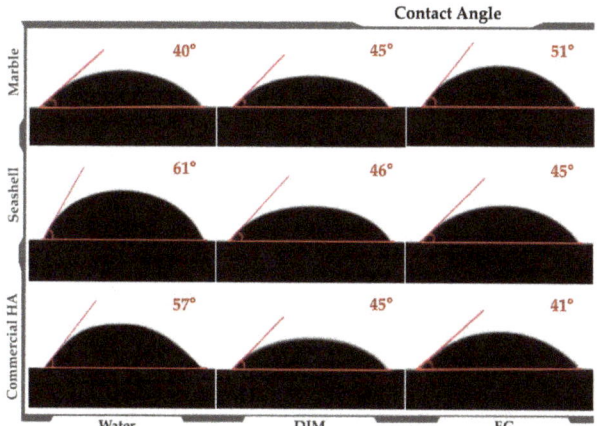

Figure 4. Contact angle assessment for marble- and seashell- derived ceramics and commercial HA samples.

3.2. Pre-Osteoblast Cell Response to Developed Ceramics

Cellular-based studies designed to evaluate the behavior of bone-derived primary cells or cell lines and mesenchymal stem cells (MSCs) represent a starting point for determining the biocompatibility of a material. In this study, we examined the viability/proliferation, adhesion and morphology, and differentiation of MC3T3-E1 pre-osteoblasts. These cells have been previously shown to exhibit stage specific genes as seen in vivo [34] and an osteoblast specific phenotype in contact with HA [35].

3.2.1. Cellular Survival and Proliferation

The capacity of the developed naturally derived calcium phosphate-based materials to support the viability and proliferation of MC3T3-E1 pre-osteoblasts plated on their surfaces has been assessed by quantifying the amount of LDH released into the culture medium, MTT reduction levels and by distinct labeling of living cells with calcein AM and of dead cells with EthD-1. Commercial HA was used as a reference material in our experiments since previous studies showed that this material belonging to the calcium phosphate family exhibited excellent biocompatibility [36,37], osteoconductivity [37–39], and osteointegration [38] abilities. Furthermore, animal and clinical studies demonstrated its direct incorporation into bone [40] and physicochemical bonding without intervening connective tissue [37].

In a first set of experiments, the possible cytotoxicity effects exerted by marble- and seashell-derived materials in comparison to commercial HA were evaluated by LDH assay after 1- and 4-days of culture. LDH is a cytosolic enzyme that can be rapidly released into the cell culture medium upon damage of the plasma membrane resulting in an increase in the OD value. As noted in Figure 5a, low absorbance values were displayed by the cells grown in contact with all three analyzed ceramics suggesting that none of them exerted cytotoxic effects. Moreover, no significant differences between the samples were observed at both incubation times indicating that they equally sustain cellular survival. In addition, the results of the MTT assay (Figure 5b) demonstrate an upward trend in OD values from 1 day to 4 days post-seeding without any significant difference between cellular substrates. Therefore, the synthesized calcium phosphate-based materials elicited an increased proliferation potential of MC3T3-E1 pre-osteoblasts. This increase appeared more obvious for marble- and seashell-derived ceramics ($p < 0.01$) than for the reference material ($p < 0.05$) although no significant differences between the three analyzed materials were noticed.

Figure 5. Cellular survival and proliferation of MC3T3-E1 pre-osteoblasts grown on tested substrates for 1- and 4-day culture periods. (a) LDH quantification as a measure of the ceramics' cytotoxicity ($n = 3$, mean ± SD); (b) MTT results showing the proliferation of cells cultured on 0.785 cm^2 of commercial HA, marble- and seashell-derived HA ($n = 3$, mean ± SD). ★ $p < 0.05$ for commercial HA, 4 day-culture vs. 1 day-culture, ★★ $p < 0.01$ for marble-derived ceramic, 4 day-culture vs. 1 day-culture, ★★ $p < 0.01$ for seashell-derived ceramic, 4 day-culture vs. 1 day-culture.

To finally assess the capacity of the developed ceramic materials to support pre-osteoblast viability and proliferation, the results of the LDH and MTT assays have been combined with the qualitative evaluation of MC3T3-E1 cells' viability and densities by performing a LIVE/DEAD Cell Viability/Cytotoxicity assay. As shown in Figure 6, the pre-osteoblasts grown in contact with all three ceramics converted non-fluorescent calcein AM to green-fluorescent calcein, revealing a high percentage of viable cells. Thus, no red fluorescent dead cells and an increasing number of green fluorescent viable cells could be observed along the culture period. These findings are in agreement with the results of the LDH and MTT assays and, collectively, suggest that marble- and seashell-derived ceramics exhibited good cytocompatibility and promoted the proliferation of pre-osteoblast cells to a similar extent with the reference material.

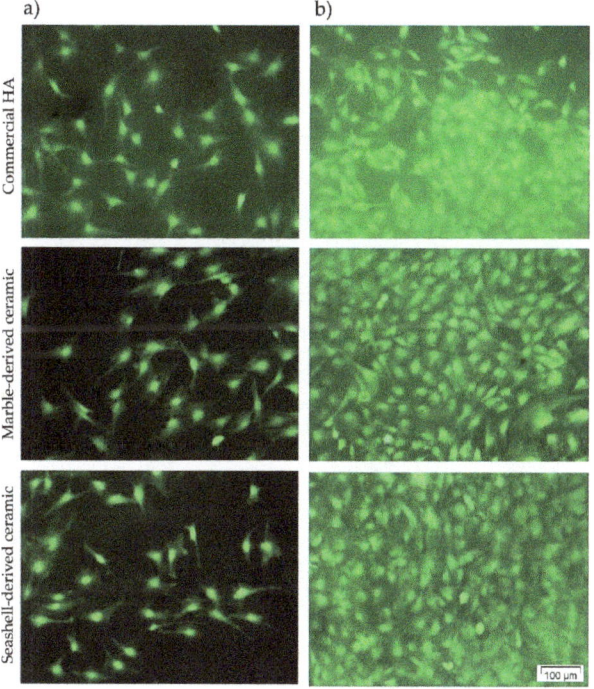

Figure 6. Fluorescence microscopy images of the MC3T3-E1 cells grown in contact with commercial HA and marble- and seashell-derived ceramics for 1 day (**a**) and 4 days (**b**). The cells were stained with a LIVE/DEAD Cell Viability/Cytotoxicity Assay Kit (live cells fluorescence green; no red fluorescent dead cells are present). Scale bar: 100 μm.

3.2.2. MC3T3-E1 Cell Adhesion and Morphological Features

For an improved understanding of the cellular interactions with the analyzed calcium phosphate-based materials, the cell adhesion and morphology were microscopically investigated after double fluorescent staining of vinculin and actin filaments. Figure 7 and Figure S1 (Supplementary Material) show the morphological features of MC3T3-E1 pre-osteoblasts on these surfaces at 3 h and 24 h post-seeding. Three hours following cell seeding, all pre-osteoblasts were attached to the samples' surfaces, and started to spread extending lamellipodia and filopodia in multiple directions (Figure 7a and Figure S1a). At the level of these cellular protrusions, discrete green-fluorescent vinculin signals can be seen on all analyzed substrates. Vinculin is an intracellular protein present in cadherin-mediated cell junctions and in focal adhesions playing a key role in initiating and establishing cell adhesion and cytoskeletal development [41]. It is worth mentioning that these immunoreactive signals are numerous on the marble- and seashell-derived ceramics suggesting the formation of focal adhesion contacts on these surfaces. Almost similar behavior was observed for commercial HA. As is well known, cell adhesion represents a cellular process accompanied by the rearrangement of cytoskeletal proteins, formation of tight focal adhesion contacts, activation of focal adhesion kinase

(FAK) and induction of various intracellular signal transduction pathways that regulate cell survival, proliferation and differentiation [42,43]. Hence, we can conclude that the two newly synthesized bioceramics are as effective as an HA surface in promoting cell adhesion and establishing tight interactions with MC3T3-E1 pre-osteoblasts. At the 24 h time point (Figure 7b and Figure S1b), well-spread cell morphologies were displayed on all surfaces, but the pre-osteoblasts attached to HA surface exhibited more stretched and elongated shapes and more discrete punctiform vinculin signals than on the marble-derived and, especially, the seashell-derived ceramics (Figure S1b). Likewise, fluorescence images showed a circumferential localization of actin filaments near the cell membrane at 3 h after cell plating (Figure 7a and Figure S1a) and well-defined thin stress fibers in the cell body at 24 h post-seeding (Figure 7b and Figure S1b) on all studied surfaces suggesting that they promote maturation of the actin cytoskeleton.

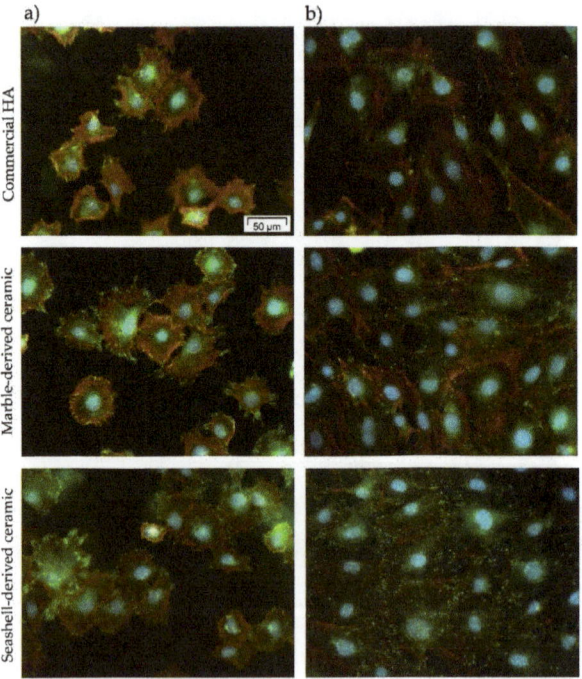

Figure 7. Merged fluorescence micrographs of actin cytoskeleton (red) and vinculin immunoreactive sites (green) in MC3T3-E1 pre-osteoblasts grown on commercial HA, marble- and seashell-derived ceramics for 3 h (**a**) and 24 h (**b**). The nuclei are labeled with DAPI (blue). Scale bar: 50 µm.

3.2.3. Pre-Osteoblast Differentiation Potential

The biological performance of the developed calcium phosphate-based materials was also evaluated by studying their potential to promote pre-osteoblast differentiation both in the absence (−OM) and presence (+OM) of the osteogenic medium containing β-GP and AA. The osteoconductive and osteoinductive effects of calcium phosphate ceramics are well documented [37–39,44] although the mechanisms through which cell differentiation is mediated are still incompletely known. In this context, the present manuscript sought to reveal the influence of the synthesized ceramics on the activity of intracellular ALP and collagen synthesis and deposition on the cell supporting materials.

As seen in Figure 8a, ALP activity of MC3T3-E1 cells grown on all analyzed ceramics continuously increased over the incubation period. After 7 days of culture, ALP activity was low whereas at 14 days post-seeding enhanced values were noticed for each cellular substrate in both experimental conditions. It is worthy to note that for all analyzed ceramic substrates, ALP activity was higher for pre-osteoblasts grown in osteogenic culture conditions versus standard conditions. Moreover, no significant differences occurred between samples except for the high enhancement of ALP activity in cells grown for 14 days under osteogenic conditions in contact with seashell-derived ceramic as compared to reference material and marble-derived ceramic ($p < 0.001$). ALP is considered an early marker of osteoblast differentiation involved in bone calcification. Specifically, ALP expression reaches a maximum level during the phase of matrix maturation, just before the onset of bone mineralization [45]. Therefore, based on the results obtained it can be concluded that marble- and seashell-ceramics are as effective as reference biomaterial in inducing matrix mineralization. However, a noticeable phenomenon was that, under osteogenic culture conditions, the ALP activity of MC3T3-E1 pre-osteoblasts was significantly higher ($p < 0.001$) on the seashell-derived ceramic than on the other analyzed calcium phosphate based-materials.

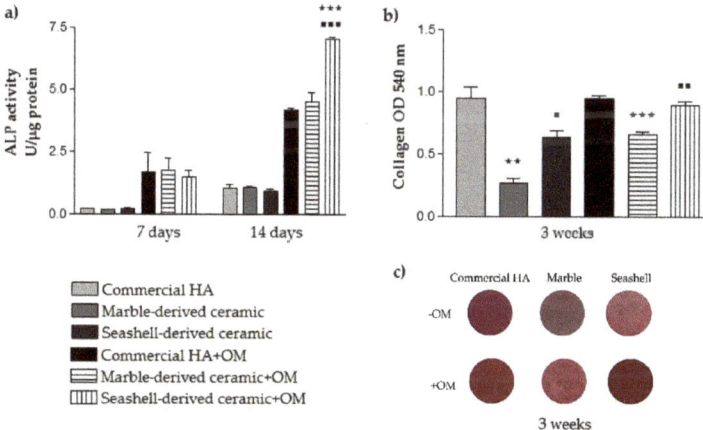

Figure 8. Results of the MC3T3-E1 cell differentiation assays showing: (**a**) the levels of intracellular ALP activity at 7 and 14 days post-seeding ($n = 3$, mean ± SD). ★★★ $p < 0.001$ for seashell-derived ceramic vs. commercial HA at 14 days post-seeding (+OM); ▪▪▪ $p < 0.001$ for seashell-derived ceramic vs. marble-derived ceramic at 14 days post-seeding (+OM); (**b**) Spectrophotometric quantification of the collagens deposited on the ceramic substrates (based on Sirius Red staining) ($n = 3$, mean ± SD). ★★ $p < 0.01$ for marble-derived ceramic vs. commercial HA (−OM); ★★★ $p < 0.001$ for marble-derived ceramic vs. commercial HA (+OM); ▪ $p < 0.05$ for seashell-derived ceramic vs. marble-derived ceramic (−OM); ▪▪ $p < 0.01$ for seashell-derived ceramic vs. marble-derived ceramic (+OM); (**c**) Digital images of the analyzed samples showing the staining of the collagenous matrix with Sirius Red.

To further establish osteogenic commitment of MC3T3-E1 pre-osteoblasts, the collagenous matrix deposited on the analyzed materials was qualitatively and quantitatively measured by Sirius Red staining. Collagens (mainly type I collagen) are the major proteins which are related to bone extracellular matrix formation. They are synthesized and secreted during the initial period of proliferation and extracellular matrix biosynthesis. Annaz et al. [46] showed that the osteoblasts are intimately connected to calcium phosphates owing to the production of extracellular collagen firmly attached to their surfaces. Our results (Figure 8b,c) showed that the amount of collagen deposited on the seashell-derived ceramic after 3 weeks of culture was similar to that expressed on the reference material in the osteogenic culture conditions. However, in standard culture conditions, less collagen ($p < 0.05$) was quantified on the seashell-derived ceramic. It is worth mentioning that independent of culture conditions, marble-derived ceramic exhibited lower amounts of collagen in comparison to the reference material and the seashell-derived ceramics. Overall, all experiments performed in this study revealed that marble-derived ceramic elicited the less favorable pre-osteoblast response in terms of bone tissue integration whereas seashell-derived ceramic exhibited almost a similar pre-osteoblast response to commercial HA. Therefore, although both developed naturally derived calcium phosphate-based materials proved to be biocompatible, seashell-derived ceramic shows great promise for substituting commercial HA in the reconstructive orthopedic field.

We assume that a surface characteristic responsible for this cellular behavior is wettability with the marble-derived ceramic displaying a water contact angle (CA) of 40° that denotes a more hydrophilic surface than seashell-derived ceramic (CA of 57°) and commercial HA (61°). It is a general trend that the cells exhibit better interaction with moderately hydrophilic substrates than hydrophobic or very hydrophilic ones [47]. Kim et al. [48] showed that the optimal wettability for cell adhesion that strongly influences subsequent cell behavior is a CA range of 50–60°. This can explain the similar and even better MC3T3-E1 pre-osteoblast response elicited by commercial HA and seashell-derived ceramic exhibiting CA values within this range.

4. Conclusions

In this work, the biological in vitro performance of recently developed naturally derived calcium phosphate-based materials, namely marble- and seashell-derived ceramics were comparatively assayed and related to commercial HA material.

Results summarize on the one hand, the potential of the adapted Rathje method for naturally derived calcium phosphates synthesis, even from the use of marble precursors, and on the other hand, the importance of thermal treatment for modulated, enhanced morpho-compositional and surface features.

Further, our studies confirmed the cytocompatibility, osteoconductivity and osteoinductivity of the commercial HA material and revealed that the newly synthesized seashell-derived ceramic exhibited a slightly better MC3T3-E1 pre-osteoblast response.

It was demonstrated that there were no distinguishable differences in the survival/ proliferation rates and adhesion/morphological features of the cells attached to the two developed naturally derived calcium-phosphate-based materials. However, seashell-derived ceramic demonstrated a higher efficacy in inducing pre-osteoblast differentiation which exhibits the potential applications these may have in the reconstructive orthopaedic field.

Supplementary Materials: The following are available online at http://www.mdpi.com/1996-1944/11/7/1097/s1: Figure S1: Fluorescent images of MC3T3-E1 pre-osteoblasts grown on commercial HA, marble- and seashell-derived ceramics for 3 h (a) and 24 h (b) respectively. Red fluorescence: actin cytoskeleton; Green fluorescence: vinculin signals. Scale bar: 50 µm.

Author Contributions: Conceptualization by A.C., F.M.; Methodology, A.C., V.M., F.M., R.I., G.E.S., I.V.A.; Software by F.M., G.E.S., I.V.A.; Validation by A.C., F.M., R.I., G.E.S., I.V.A.; Formal Analysis by V.M., F.M.; Investigation by V.M., R.I., F.M., G.E.S., I.V.A.; Resources by A.C., V.M., R.I., M.G.N., F.M., A.-C.M., G.E.S., I.V.A.; Data Curation by A.C., F.M.; Writing-Original Draft Preparation by V.M., R.I., M.G.N., A.-C.M.; Writing-Review & Editing, A.C., F.M., A.-C.M.; Visualization by A.C., V.M., R.I., M.G.N., A.-C.M.; Supervision by A.C., F.M.; Project Administration by A.C., F.M.; Funding Acquisition by A.C., F.M.

Funding: This research was funded by the Romanian Ministry of National Education, CNCS-UEFISCDI, grant PED 108/2017.

Conflicts of Interest: The authors declare no conflict of interest.

References

1. Roy, D.M.; Linnehan, S.K. Hydroxyapatite formed from coral skeletal carbonate by hydrothermal exchange. *Nature* **1974**, *247*, 220–222. [CrossRef] [PubMed]
2. Miculescu, F.; Mocanu, A.C.; Maidaniuc, A.; Dascalu, C.A.; Miculescu, M.; Voicu, S.; Ciocoiu, R.C. Biomimetic calcium phosphates derived from marine and land bioresources. In *Hydroxyapatite—Advances in Composite Nanomaterials, Biomedical Applications and Its Technological Facets*; InTechOpen: London, UK, 2018; pp. 89–108.

3. Miculescu, F.; Mocanu, A.C.; Stan, G.E.; Miculescu, M.; Maidaniuc, A.; Cîmpean, A.; Mitran, V.; Voicu, S.I.; Machedon-Pisu, T.; Ciocan, L.T. Influence of the modulated two-step synthesis of biogenic hydroxyapatite on biomimetic products' surface. *Appl. Surf. Sci.* **2017**, *438*, 147–157. [CrossRef]
4. Pujiyanto, E.; Widyo Laksono, P.; Triyono, J. Synthesis and characterization of hydroxyapatite powder from natural gypsum rock. *Adv. Mater. Res.* **2014**, *893*, 56–59. [CrossRef]
5. Wu, S.-C.; Hsu, H.-C.; Wu, Y.-N.; Ho, W.-F. Hydroxyapatite synthesized from oyster shell powders by ball milling and heat treatment. *Mater. Charact.* **2011**, *62*, 1180–1187. [CrossRef]
6. Miculescu, F.; Stan, G.; Ciocan, L.; Miculescu, M.; Berbecaru, A.; Antoniac, I. Cortical bone as resource for producing biomimetic materials for clinical use. *Dig. J. Nanomater. Biostruct.* **2012**, *7*, 1667–1677.
7. Pal, A.; Maity, S.; Chabri, S.; Bera, S.; Chowdhury, A.R.; Das, M.; Sinha, A. Mechanochemical synthesis of nanocrystalline hydroxyapatite from mercenaria clam shells and phosphoric acid. *Biomed. Phys. Eng. Express* **2017**, *3*. [CrossRef]
8. Bouler, J.; Pilet, P.; Gauthier, O.; Verron, E. Biphasic calcium phosphate ceramics for bone reconstruction: A review of biological response. *Acta Biomater.* **2017**, *53*, 1–12. [CrossRef] [PubMed]
9. Denry, I.; Kuhn, L.T. Design and characterization of calcium phosphate ceramic scaffolds for bone tissue engineering. *Dental Mater.* **2016**, *32*, 43–53. [CrossRef] [PubMed]
10. Wang, P.; Zhao, L.; Liu, J.; Weir, M.D.; Zhou, X.; Xu, H.H. Bone tissue engineering via nanostructured calcium phosphate biomaterials and stem cells. *Bone Res.* **2014**, *2*. [CrossRef] [PubMed]
11. Sadowska, J.-M.; Guillem-Marti, J.; Montufar, E.B.; Espanol, M.; Ginebra, M.-P. Biomimetic versus sintered calcium phosphates: The in vitro behavior of osteoblasts and mesenchymal stem cells. *Tissue Eng. Part A* **2017**, *23*, 1297–1309. [CrossRef] [PubMed]
12. Joschek, S.; Nies, B.; Krotz, R.; Göpferich, A. Chemical and physicochemical characterization of porous hydroxyapatite ceramics made of natural bone. *Biomaterials* **2000**, *21*, 1645–1658. [CrossRef]
13. Cozza, N.; Monte, F.; Bonani, W.; Aswath, P.; Motta, A.; Migliaresi, C. Bioactivity and mineralization of natural hydroxyapatite from cuttlefish bone and bioglass® co-sintered bioceramics. *J. Tissue Eng. Regen. Med.* **2018**, *12*, e1131–e1142. [CrossRef] [PubMed]
14. Akram, M.; Ahmed, R.; Shakir, I.; Ibrahim, W.A.W.; Hussain, R. Extracting hydroxyapatite and its precursors from natural resources. *J. Mater. Sci.* **2014**, *49*, 1461–1475. [CrossRef]
15. Abdulrahman, I.; Tijani, H.I.; Mohammed, B.A.; Saidu, H.; Yusuf, H.; Ndejiko Jibrin, M.; Mohammed, S. From garbage to biomaterials: An overview on egg shell based hydroxyapatite. *J. Mater.* **2014**. [CrossRef]
16. Łukaszewska-Kuska, M.; Krawczyk, P.; Martyla, A.; Hędzelek, W.; Dorocka-Bobkowska, B. Hydroxyapatite coating on titanium endosseous implants for improved osseointegration: Physical and chemical considerations. *Adv. Clin. Exp. Med.* **2018**. [CrossRef]
17. Ohe, M.; Moridaira, H.; Inami, S.; Takeuchi, D.; Nohara, Y.; Taneichi, H. Pedicle screws with a thin hydroxyapatite coating for improving fixation at the bone-implant interface in the osteoporotic spine: Experimental study in a porcine model. *J. Neurosurg. Spine* **2018**, *28*, 679–687. [CrossRef] [PubMed]
18. Hing, K.A. Bone repair in the twenty-first century: Biology, chemistry or engineering? *Philos. Trans. Ser. A Math. Phys. Eng. Sci.* **2004**, *362*, 2821–2850. [CrossRef] [PubMed]
19. Hong, Z.; Zhang, P.; He, C.; Qiu, X.; Liu, A.; Chen, L.; Chen, X.; Jing, X. Nano-composite of poly (l-lactide) and surface grafted hydroxyapatite: Mechanical properties and biocompatibility. *Biomaterials* **2005**, *26*, 6296–6304. [CrossRef] [PubMed]
20. Janus, A.M.; Faryna, M.; Haberko, K.; Rakowska, A.; Panz, T. Chemical and microstructural characterization of natural hydroxyapatite derived from pig bones. *Microchim. Acta* **2008**, *161*, 349–353. [CrossRef]
21. Swetha, M.; Sahithi, K.; Moorthi, A.; Srinivasan, N.; Ramasamy, K.; Selvamurugan, N. Biocomposites containing natural polymers and hydroxyapatite for bone tissue engineering. *Int. J. Biol. Macromol.* **2010**, *47*, 1–4. [CrossRef] [PubMed]
22. Yoshikawa, H.; Myoui, A. Bone tissue engineering with porous hydroxyapatite ceramics. *J. Artif. Organs* **2005**, *8*, 131–136. [CrossRef] [PubMed]
23. Kubasiewicz-Ross, P.; Hadzik, J.; Seeliger, J.; Kozak, K.; Jurczyszyn, K.; Gerber, H.; Dominiak, M.; Kunert-Keil, C. New nano-hydroxyapatite in bone defect regeneration: A histological study in rats. *Ann. Anat.* **2017**, *213*, 83–90. [CrossRef] [PubMed]
24. Vecchio, K.S.; Zhang, X.; Massie, J.B.; Wang, M.; Kim, C.W. Conversion of bulk seashells to biocompatible hydroxyapatite for bone implants. *Acta Biomater.* **2007**, *3*, 910–918. [CrossRef] [PubMed]

25. Tripathi, G.; Basu, B. A porous hydroxyapatite scaffold for bone tissue engineering: Physico-mechanical and biological evaluations. *Ceram. Int.* **2012**, *38*, 341–349. [CrossRef]
26. Rathje, W. Zur kenntnis der phosphate i: Über hydroxylapatit. *Bodenkd. Pflanzenernähr.* **1939**, *12*, 121–128. [CrossRef]
27. Miculescu, F.; Mocanu, A.-C.; Dascălu, C.A.; Maidaniuc, A.; Batalu, D.; Berbecaru, A.; Voicu, S.I.; Miculescu, M.; Thakur, V.K.; Ciocan, L.T. Facile synthesis and characterization of hydroxyapatite particles for high value nanocomposites and biomaterials. *Vacuum* **2017**, *146*, 614–622. [CrossRef]
28. Maidaniuc, A.; Miculescu, F.; Voicu, S.I.; Andronescu, C.; Miculescu, M.; Matei, E.; Mocanu, A.C.; Pencea, I.; Csaki, I.; Machedon-Pisu, T. Induced wettability and surface-volume correlation of composition for bovine bone derived hydroxyapatite particles. *Appl. Surf. Sci.* **2017**, *438*, 158–166. [CrossRef]
29. Maidaniuc, A.; Miculescu, M.; Voicu, S.; Ciocan, L.; Niculescu, M.; Corobea, M.; Rada, M.; Miculescu, F. Effect of micron sized silver particles concentration on the adhesion induced by sintering and antibacterial properties of hydroxyapatite microcomposites. *J. Adhes. Sci. Technol.* **2016**, *30*, 1829–1841. [CrossRef]
30. Miculescu, F.; Maidaniuc, A.; Voicu, S.I.; Thakur, V.K.; Stan, G.E.; Ciocan, L. Progress in hydroxyapatite-starch based sustainable biomaterials for biomedical bone substitution applications. *ACS Sustain. Chem. Eng.* **2017**, *5*, 8491–8512. [CrossRef]
31. Neacsu, P.; Staras, A.I.; Voicu, S.I.; Ionascu, I.; Soare, T.; Uzun, S.; Cojocaru, V.D.; Pandele, A.M.; Croitoru, S.M.; Miculescu, F. Characterization and in vitro and in vivo assessment of a novel cellulose acetate-coated mg-based alloy for orthopedic applications. *Materials* **2017**, *10*, 686. [CrossRef] [PubMed]
32. Gordin, D.; Ion, R.; Vasilescu, C.; Drob, S.; Cimpean, A.; Gloriant, T. Potentiality of the "gum metal" titanium-based alloy for biomedical applications. *Mater. Sci. Eng. C* **2014**, *44*, 362–370. [CrossRef] [PubMed]
33. Boyan, B.D.; Hummert, T.W.; Dean, D.D.; Schwartz, Z. Role of material surfaces in regulating bone and cartilage cell response. *Biomaterials* **1996**, *17*, 137–146. [CrossRef]
34. Quarles, L.D.; Yohay, D.A.; Lever, L.W.; Caton, R.; Wenstrup, R.J. Distinct proliferative and differentiated stages of murine mc3t3-e1 cells in culture: An in vitro model of osteoblast development. *J. Bone Miner. Res.* **1992**, *7*, 683–692. [CrossRef] [PubMed]
35. Matsumoto, T.; Kawakami, M.; Kuribayashi, K.; Takenaka, T.; Minamide, A.; Tamaki, T. Effects of sintered bovine bone on cell proliferation, collagen synthesis, and osteoblastic expression in mc3t3-e1 osteoblast-like cells. *J. Oorthop. Res.* **1999**, *17*, 586–592. [CrossRef] [PubMed]
36. Aktuğ, S.L.; Durdu, S.; Yalçın, E.; Çavuşoğlu, K.; Usta, M. Bioactivity and biocompatibility of hydroxyapatite-based bioceramic coatings on zirconium by plasma electrolytic oxidation. *Mater. Sci. Eng. C* **2017**, *71*, 1020–1027. [CrossRef] [PubMed]
37. Jarcho, M. Calcium phosphate ceramics as hard tissue prosthetics. *Clin. Orthop. Relat. Res.* **1981**, *157*, 259–278. [CrossRef]
38. Jaramillo, C.D.; Rivera, J.A.; Echavarría, A.; O'byrne, J.; Congote, D.; Restrepo, L.F. Osteoconductive and osseointegration properties of a commercial hydroxyapatite compared to a synthetic product. *Rev. Colomb. Cienc. Pecu.* **2010**, *23*, 471–483.
39. Do Prado Ribeiro, D.C.; de Abreu Figueira, L.; Mardegan Issa, J.P.; Dias Vecina, C.A.; JoséDias, F.; Da Cunha, M.R. Study of the osteoconductive capacity of hydroxyapatite implanted into the femur of ovariectomized rats. *Microsc. Res. Tech.* **2012**, *75*, 133–137. [CrossRef] [PubMed]
40. Denissen, H.; De Groot, K.; Makkes, P.C.; Van den Hooff, A.; Klopper, P. Tissue response to dense apatite implants in rats. *J. Biomed. Mater. Res. Part A* **1980**, *14*, 713–721. [CrossRef] [PubMed]
41. Carisey, A.; Ballestrem, C. Vinculin, an adapter protein in control of cell adhesion signalling. *Eur. J. Cell Biol.* **2011**, *90*, 157–163. [CrossRef] [PubMed]
42. Anselme, K. Osteoblast adhesion on biomaterials. *Biomaterials* **2000**, *21*, 667–681. [CrossRef]
43. Wozniak, M.A.; Modzelewska, K.; Kwong, L.; Keely, P.J. Focal adhesion regulation of cell behavior. *Biochim. Biophys. Acta* **2004**, *1692*, 103–119. [CrossRef] [PubMed]
44. Lee, K.; Park, M.; Kim, H.; Lim, Y.; Chun, H.; Kim, H.; Moon, S. Ceramic bioactivity: Progresses, challenges and perspectives. *Biomed. Mater.* **2006**, *1*. [CrossRef] [PubMed]
45. Boyan, B.D.; Lohmann, C.H.; Dean, D.D.; Sylvia, V.L.; Cochran, D.L.; Schwartz, Z. Mechanisms involved in osteoblast response to implant surface morphology. *Annu. Rev. Mater. Res.* **2001**, *31*, 357–371. [CrossRef]
46. Annaz, B.; Hing, K.; Kayser, M.; Buckland, T.; Di Silvio, L. An ultrastructural study of cellular response to variation in porosity in phase-pure hydroxyapatite. *J. Microsc.* **2004**, *216*, 97–109. [CrossRef] [PubMed]

47. Arima, Y.; Iwata, H. Effect of wettability and surface functional groups on protein adsorption and cell adhesion using well-defined mixed self-assembled monolayers. *Biomaterials* **2007**, *28*, 3074–3082. [CrossRef] [PubMed]
48. Kim, S.H.; Ha, H.J.; Ko, Y.K.; Yoon, S.J.; Rhee, J.M.; Kim, M.S.; Lee, H.B.; Khang, G. Correlation of proliferation, morphology and biological responses of fibroblasts on LDPE with different surface wettability. *J. Biomater. Sci. Polym. Ed.* **2007**, *18*, 609–622. [CrossRef] [PubMed]

© 2018 by the authors. Licensee MDPI, Basel, Switzerland. This article is an open access article distributed under the terms and conditions of the Creative Commons Attribution (CC BY) license (http://creativecommons.org/licenses/by/4.0/).

Article

Nanoscale Electrical Potential and Roughness of a Calcium Phosphate Surface Promotes the Osteogenic Phenotype of Stromal Cells

Igor A. Khlusov [1,2,*], Yuri Dekhtyar [3], Yurii P. Sharkeev [4,5], Vladimir F. Pichugin [4], Marina Y. Khlusova [6], Nataliya Polyaka [3], Fjodors Tjulkins [3], Viktorija Vendinya [3], Elena V. Legostaeva [5], Larisa S. Litvinova [2], Valeria V. Shupletsova [2], Olga G. Khaziakhmatova [2], Kristina A. Yurova [2] and Konstantin A. Prosolov [4,5]

1. Research School of Chemistry & Applied Biomedical Sciences, National Research Tomsk Polytechnic University, Tomsk 634050, Russia
2. Basic Laboratory of Immunology and Cell Biotechnology, Immanuel Kant Baltic Federal University, Kaliningrad 236041, Russia; larisalitvinova@yandex.ru (L.S.L.); vshupletsova@mail.ru (V.V.S.); hazik36@mail.ru (O.G.K.); kristina_kofanova@mail.ru (K.A.Y.)
3. Institute of Biomedical Engineering and Nanotechnologies, Riga Technical University, Riga LV-1658, Latvia; dekhtyar@latnet.lv (Y.D.); natalija.polaka@latnet.lv (N.P.); Kgam@inbox.lv (F.T.); vicken@inbox.lv (V.V.)
4. Research School of High-Energy Physics, National Research Tomsk Polytechnic University, Tomsk 634050, Russia; sharkeev@ispms.tsc.ru (Y.P.S.); pichugin@tpu.ru (V.F.P.); konstprosolov@gmail.com (K.A.P.)
5. Institute of Strength Physics and Materials Science of SB RAS, Tomsk 634055, Russia; lego@ispms.tsc.ru
6. Department of Pathophysiology, Siberian State Medical University, Tomsk 634050, Russia; uchsovet@ssmu.ru
* Correspondence: khlusov63@mail.ru; Tel.: +8-3822-901-101 (ext. 1823)

Received: 25 April 2018; Accepted: 7 June 2018; Published: 9 June 2018

Abstract: Mesenchymal stem cells (MSCs) and osteoblasts respond to the surface electrical charge and topography of biomaterials. This work focuses on the connection between the roughness of calcium phosphate (CP) surfaces and their electrical potential (EP) at the micro- and nanoscales and the possible role of these parameters in jointly affecting human MSC osteogenic differentiation and maturation in vitro. A microarc CP coating was deposited on titanium substrates and characterized at the micro- and nanoscale. Human adult adipose-derived MSCs (hAMSCs) or prenatal stromal cells from the human lung (HLPSCs) were cultured on the CP surface to estimate MSC behavior. The roughness, nonuniform charge polarity, and EP of CP microarc coatings on a titanium substrate were shown to affect the osteogenic differentiation and maturation of hAMSCs and HLPSCs in vitro. The surface EP induced by the negative charge increased with increasing surface roughness at the microscale. The surface relief at the nanoscale had an impact on the sign of the EP. Negative electrical charges were mainly located within the micro- and nanosockets of the coating surface, whereas positive charges were detected predominantly at the nanorelief peaks. HLPSCs located in the sockets of the CP surface expressed the osteoblastic markers osteocalcin and alkaline phosphatase. The CP multilevel topography induced charge polarity and an EP and overall promoted the osteoblast phenotype of HLPSCs. The negative sign of the EP and its magnitude at the micro- and nanosockets might be sensitive factors that can trigger osteoblastic differentiation and maturation of human stromal cells.

Keywords: surface roughness; surface electrical potential; micro- and nanoscale; human mesenchymal cells; osteocalcin; alkaline phosphatase; in vitro

1. Introduction

The biological hierarchy from the nanoscale (molecules) to the macrodimensional (organs and organisms) via microsized cells and mesosized tissues clearly indicates that a nano-to-meso-to-macro multilevel approach should be employed for the engineering of biomaterials. Nanoscale engineering is a platform for the "bottom-to-top" approach to the design of biomaterials because organisms communicate with biomaterials at the biomolecule–biomaterial interface.

Modern biomaterial biocompatibility research focuses on stem cell (SC)–biomaterial interactions (that occur at the nano- and microscales) because SCs are the fundamental units that produce/regenerate tissue.

To control SCs, biomaterials mimic the bioimplant surface (BS), with both its morphology and physical/chemical properties being engineered. Only cells adhered to the BS are involved in tissue regeneration when the implant serves as a scaffold.

The influence of surface morphology/roughness on cell attachment has been studied intensively. The significant increase in the number of publications in this area started in 1990, and the number of these publications reached approximately 1100 by April 2017 (SCOPUS, keywords implant-surface-roughness-cell), which is equivalent to 40–55 publications annually.

The physical and chemical properties of a surface are identified by fundamental factors, such as the surface energy (SE), which strongly influences the attachment of cells. The cells adhere to the BS via a specific protein layer that coats the implant shortly after its insertion into a living organism [1]. Fundamentally, attachment of a specific protein molecule obeys adhesion theory [2], which considers dispersion and electrostatic interactions between the particle and the substrate. As the dispersive interaction potential decreases much more quickly with distance r ($\sim r^{-6}$) [3] than the electrostatic potential ($\sim r^{-1}$), the the electrostatic potential is expected to have a stronger impact on trapping an electrically charged "tail" of a molecule that travels at some distance from the substrate surface. The electrostatic interaction could be promoted by engineering the electrical charge at the BS. Such an approach is available for dielectric biomaterials, and hydroxyapatite (HAP, a very popular material for implant fabrication) is among them.

The influence of the electrostatic factor on cell attachment has been less explored than the influence of surface morphology/roughness. A significant increase in the number of publications on this topic also started in 1990, and the total number reached approximately 270 by April 2017 (SCOPUS, keywords implant-surface-charge-cell), which is equivalent to 10–13 publications annually.

A few biomaterials are currently in use. However, the present article focuses on the widely exploited biomaterial calcium phosphate (CP). The surface layer electrical charge of CP can be engineered by radiation [4], electrical polarization [5], doping, and reconstruction of the surface atomic/molecular coupling [6]. The surface electrical charge of CP nanoparticles, which are sometimes used to construct an implant, depends on the particle radii [6]. In this connection, the surface roughness, having peaks and valleys that are characterized by the radii of its building blocks, could influence the surface charge.

The surface morphology of CP is typically engineered because there is a set of methods for forming CP coatings, such as sol-gel, plasma spraying, magnetron sputtering, and detonation gas spraying methods. However, these methods have a number of disadvantages. One of the most significant disadvantages is the absence of chemical bonds between the formed coating and the substrate surface. Moreover, the process is long, and the cost of the final product is high.

The abovementioned methods are "direct vision" methods; therefore, they are not used for applying CP coatings to implants. Microarc oxidation (MAO) is more convenient and effective for producing CP coatings on implants with complex shapes and produces good physical and chemical properties [7,8].

The electrolyte (components, pH, and temperature) and the electrophysical parameters of microarc oxidation, such as the anode and anode-cathode modes, voltage, current density, pulse period frequency, time period, and material type (treatment, surface roughness, and electroconductivity) affect the physical, chemical, and biological characteristics of the coating.

A set of key surface physicochemical parameters that affect cells has already been identified, namely, surface mechanical integrity, the rate of scaffold degradation, fluid transport, cell-recognizable surface chemical properties, the ability to induce signal transduction [9], surface free energy, wettability [10], surface electrical charge [11] and chemical composition [12], and surface structure, morphology, and stiffness [13,14].

There is evidence in favor of the influence of HAP surface charge on interactions with biomolecules and cells. The electrical charge deposited on a HAP surface alters its wettability and influences the absorption of proteins [4]. Research reported by [15] demonstrated that a HAP surface electrical potential (EP) shift of 0.5 V influences osteoblastic cell attachment.

In 1964, Curtis and Varde [16] assumed that surface topography has the most important effect on cells. Nevertheless, scientists are only just beginning to understand niche–cell interactions [17]. The achieved results are not unequivocal, and different conditions for fabricating scaffoldings have been recognized as possible factors for the varying outcomes [18]. A series of works [19–22] focused on determination of the relationship between metallic surface roughness and cell-adhesion parameters. No relation was found between the roughness amplitude, R_a, and cell adhesion. The parameter that better discriminates adhesion is the mean distance between asperities (S_m); however, this correlation is insufficiently strict, and several authors [22–24] have developed a new parameter called "adhesion power".

A wide range of cells (fibroblasts, osteoblasts, osteoclasts, nerve cells, stromal SCs, and embryonic SCs, among others) respond to surface micro- [25–28] and nanotopography [11,17,29], and the topography is the main influencing factor [12].

However, the connection between the multiple extracellular physicochemical events that control and trigger cells, particularly mesenchymal stem cells (MSCs), is still unclear. The electric field plays a crucial role in the effects of ion channels and cellular ζ potentials [30–33]. Understanding the relationship between the electric charge of the material surface and protein adsorption is a focus in examining the mechanisms of tissue integration with biomaterials [34]. MSCs and osteoblasts respond to surface charge [34–38] and micro- [39] and nanoreliefs [40,41].

The combination of Ti and CP forms a modern material that is widely used for bioimplant design. The electrical charge of CP surfaces affects their wettability [42], which results in molecular adhesion [43]. The microroughness of CP coatings deposited on Ti using the microarc technique has been shown to control mouse bone tissue remodeling in vivo [44] and osteogenic differentiation of human MSCs in vitro [28].

From the above results, both the morphology and electrical properties of a surface appear to control cells. However, from the electrostatic point of view, the morphology of the surface affects its electrical charge density. Thus, the surface morphology and its electrical charge are possibly interconnected. The nanoscale morphology and particularly the sharpness of the surface affect its electric charge [45], which has its own effect on cells.

Nevertheless, the balance between the physical events that control cell behavior is still unclear. Current studies have not connected surface micro- and nanotopography and the electrical charge as joint factors affecting MSC fate.

This work focuses on the connection between CP surface roughness and the EP at the micro- and nanoscale because they are correlated factors that affect human MSC osteogenic differentiation and maturation in vitro.

2. Materials and Methods

2.1. Substrate Preparation and CP Deposition

Commercially pure titanium (99.58 Ti, 0.12 O, 0.18 Fe, 0.07 C, 0.04 N, and 0.01 H wt %) plates (10 × 10 × 1 mm^3) were used as substrates for the deposition of CP coatings. The samples were polished with silicon-carbide paper of 120, 480, 600, and 1200 grit. Then, the samples were cleaned ultrasonically with an Elmasonic S10 device (Elma Schmidbauer GmbH, Siegen, Germany) for 10 min in distilled water immediately before deposition.

Bilateral CP coatings were prepared in the anodal regime as described previously [46] using a Micro-Arc 3.0 apparatus (ISPMS, Tomsk, Russia). The setup consisted of positive and negative pulse power sources, a computer for controlling the deposition process, a galvanic bath with a water cooling system, and electrodes. An aqueous solution prepared from 20 wt % phosphoric acid, 6 wt % dissolved HAP powder, and 9 wt % dissolved calcium carbonate was used to obtain CP. Microarc oxidation was performed in the anodic regime with initial current densities in the range of 0.2–0.25 A/cm^2. In previous work [47], the optimal microarc oxidation parameters for the deposition of a CP coating on titanium were a pulse frequency of 100 Hz, a pulse duration of 50 μs, and a deposition time of 5–10 min. Samples with different CP coating thickness and CP adhesion strength to the substrate were prepared by varying the electrical voltage in the range of 150–400 V. CP coatings with a thickness of 10–150 μm and a roughness R_a of 1.5–6.5 μm were deposited by varying the electrical voltage and deposition time.

The specimens were dried in a dry-heat manner with a Binder FD53 oven (Binder GmbH, Tuttlingen, Germany) at 453 K for 1 h.

To measure the adhesion strength of the CP coating to the substrate, two cylinders were glued with Loctite Hysol 9514 glue to both sides of the sample with the coating. They were fixed in grips for testing under tension in an Instron 1185 machine (Instron-1185, Buckinghamshire, UK). The adhesion strength was considered to be the maximum stress required to tear the cylinders of the CP coating and was determined by $\sigma = F/S$, where F is the breakout force and S is the separation area.

2.2. Surface Characterization

2.2.1. Surface and Microstructure Observation and Phase Composition

The surface topography and elemental composition of the coating surface and cells were analyzed via scanning electron microscopy (SEM, Philips SEM 515, Amsterdam, The Netherlands) using a microscope equipped with an energy-dispersive X-ray spectroscope (EDAX ECON IV). The surface area was randomly examined at magnifications ranging from 500 to 10,000×. The microstructure of CP coatings was investigated using transmission electron microscopy (TEM, FEI Tecnai 20, Hillsboro, OR, USA). The phase composition was determined by X-ray diffraction (XRD, Bruker D8 Advance, Billerica, MA, USA) in the angular range of 2θ = 5–90° with a scan step size of 0.010° and Cu K_a radiation.

2.2.2. Microroughness Measurements

The substrate surface roughness was assessed with a Talysurf 5–120 profilometer (Taylor Hobson Ltd., Leicester, UK). Measurements of roughness amplitude parameters and the mean value of the profile element width, S_m, were carried out according to the standards ISO 4287-1997 and ISO3274-1996, respectively [48]. Ten randomly selected traces were recorded for each specimen. The average roughness (R_a), peak-to-valley roughness (R_z), and maximum roughness (R_{max}) were estimated. A strong linear correlation ($r = 0.95$; significance 99%) was identified between R_a and R_z or R_{max}. Therefore, only R_a was used for further roughness characterization.

2.2.3. Optical Microscopy and SEM

An Olympus GX-71 inverted metallographic microscope (Olympus Corporation, Tokyo, Japan) equipped with an Olympus DP 70 digital camera was used to obtain dark field images of the coating relief and to locate alkaline phosphatase (ALP)- or osteocalcin (OCN)-stained cells. In addition, valleys and sockets were analyzed on the CP coatings. Magnifications of 500× and 1000× were applied. Cell morphology and size were demonstrated via SEM (LEO EVO 50, Zeiss, Oberkochen, Germany). Cells were prepared for SEM as described previously [28].

2.2.4. Atomic Force and Kelvin Probe Microscopy

Atomic force microscopy (AFM) was applied in contact mode to characterize the nanoscale roughness of the samples. The average nanoroughness, R_{an}, was measured. Kelvin probe force measurements were used to identify the surface potential at the micro/nanoscale. A Solver–PRO47 microscope (NT-MDT Co., Zelenograd, Russia) was used for both AFM and Kelvin probe force measurements.

2.3. Surface EP

2.3.1. Surface Electric Charge at the Macroscale

The method of lifting the electrode (the Eguchi method) [49] was used to measure the macroscale EP of the surface. The measurements were carried out under ambient conditions. A homemade device was used and is described in detail in [50]. The device measures the electric field potential of weakly charged bodies. The longitudinal resolution of the device was 5 mm, and the measured potentials ranged from tens of millivolts to hundreds of volts. An electrode installed on the surface of the coating was used to measure the charge. The potential induced at the measuring electrode, V_{in}, is related to the potential of the surface, V_L, by the expression:

$$V_L = \frac{C_{in} + S_l}{S_l} V_{in} \quad (1)$$

where C_{in} is the input capacitance of the measuring instrument and C_l is the measured capacitance.

2.3.2. Electron Work Function

Because the V_L measurements were performed in air, the voltage could be affected by the environment, and some results appeared dubious. To overcome these doubts, the electron work function (φ) of the specimen surface was estimated based on photoelectron emission detection under high-vacuum conditions.

The value of φ is the minimal energy required for an electron to escape from a solid. The electrical field induced by the surface electrical charge contributes to the value of φ, which increases when the charge is altered in the negative direction. An increment in φ ($\Delta\varphi$) is an index for the surface EP shift.

The specimens were irradiated by soft ultraviolet (UV) light at 3–6 eV (the range of the expected value of φ) to release an electron, and the photoelectron emission current (I) was measured. The spectrometer used is described in detail in [51]. The value of φ was identified as the energy of the photons when $I = 0$.

2.4. Biological Testing

Before biological testing, the samples were dry-heat sterilized as described above.

2.4.1. Cell Culture and In Vitro Staining

(Human cell isolation and cultivation was permitted by the Local Ethics Committee, Innovation Park, Immanuel Kant Baltic Federal University (permit no. 7 from December 9, 2015).)

Adult adipose-derived MSCs (AMSCs) were isolated from human lipoaspirates of healthy volunteers [52]. The cells were stained using a Phenotyping Kit, human (130-095-198) and Viability Fixable Dyes (Miltenyi Biotec, Bergisch Gladbach, Germany), and the results were analyzed with a MACS Quant flow cytometer (Miltenyi Biotec, Bergisch-Gladbach, Germany) and KALUZA Analysis Software (Beckman Coulter, Brea, CA, USA) according to the manufacturer's instructions. More than 99% of the viable cells expressed CD73, CD90, or CD105 markers and did not display CD45, CD34, CD20, or CD14 markers (less than 1%).

To confirm the morphofunctional nature of AMSCs, 5×10^4 cells/mL were cultivated in 1.5 mL of medium with reagent from a StemPro® Differentiation Kit (Thermo Fisher Scientific, Waltham, MA, USA) for 21 days, and the medium was exchanged every 3–4 days. After removal of the supernatants, the adherent cells were dried, fixed, and stained with the dyes. According to the recommendations of the International Society for Cellular Therapy (ISCT) and the International Federation for Adipose Therapeutics and Science (IFATS) [52], the multipotent behavior of AMSCs was estimated by cell staining with alcian blue (Sigma-Aldrich, St. Louis, MO, USA), which enabled visualization of proteoglycan synthesis by chondrocytes; alizarin red S (Sigma-Aldrich, St. Louis, MO, USA), which identified mineralization of intercellular substances by osteoblasts; and oil red (Sigma-Aldrich, St. Louis, MO, USA), which detected neutral triglycerides and lipids in adipocytes. All staining was performed as recommended by the manufacturer. The results were assessed with a Zeiss Axio Observer A1 microscope (Carl Zeiss Microscopy, LLC, Thornwood, NY, USA) using ZEN 2012 software (Carl Zeiss Microscopy, LLC).

To determine the effect of CP coatings on AMSC osteogenic differentiation, cells were cultured in an incubator (Sanyo, Japan) at 100% humidity with 5% carbon dioxide at 37 °C for 21 days (the medium was replaced with fresh medium every 3–4 days). The cell suspension was freshly prepared at a concentration of 1.5×10^5 viable cells /mL in 1.5 mL of the following culture medium: 90% Dulbecco's modified Eagle's medium (DMEM)/F12 (1:1) (Gibco Life Technologies; Grand Island, NY, USA), 10% fetal bovine serum (Sigma-Aldrich, St. Louis, MO, USA), 50 mg/L gentamicin (Invitrogen, Carlsbad, CA, USA), and sterile L-glutamine solution freshly added to a final concentration of 280 mg/L (Sigma-Aldrich). A titanium substrate with a two-sided microarc CP coating was placed in a plastic well of a 12-well (well area of 1.86 cm^2) flat-bottom plate (Orange Scientific, Braine-l'Alleud, Belgium).

To establish the self-differentiation potency of cells on a rough CP surface, the culture medium was not saturated by osteogenic supplements. The cells were cultured in DMEM/F12 culture medium either with or without the tested samples (cells were seeded on the samples and around them) and in osteogenic medium from a StemPro® Differentiation Kit (Thermo Fisher Scientific, Waltham, MA, USA). The osteoblasts in cell cultures were stained with alizarin red S as described above. Digital images of AMSCs cultured on CP coatings were obtained via reflected light microscopy on an Olympus GX-71 metallographic device (Olympus Corporation, Tokyo, Japan).).

Prenatal stromal cells from the human lung (HLPSCs) with a CD34$^-$CD44$^+$OCN$^-$ immunophenotype (Stem Cell Bank Ltd., Tomsk, Russia) were used to study MSC osteogenic differentiation and maturation induced by the nanoscale EP and the roughness of the CP coatings. HLPSCs react to the microroughness of the microarc CP surface according to our previous in vitro investigation [28]. The details are provided in [53]. After the cells were thawed, the viability (92%) of the cells was determined with an ISO 10993-5 test; 0.4% trypan blue was used. One CP-coated specimen was placed in each plastic well of a 24-well flat-bottom plate (Orange Scientific, Braine-l'Alleud, Belgium). The HLPSC suspension was freshly prepared at a concentration of 3×10^4 viable cells /mL in the following culture medium: 80% DMEM/F12 (1:1) (Gibco Life Technologies; Grand Island, NY, USA), 20% fetal bovine serum (Sigma-Aldrich, St. Louis, MO, USA), 50 mg/L gentamicin (Invitrogen, UK), and L-glutamine sterile solution freshly added to a final concentration of 280 mg/L (Sigma-Aldrich). To determine the osteogenic potency of a rough CP surface, the culture medium did not contain osteogenic supplements, such as β-glycerophosphate, dexamethasone, and ascorbic acid.

The cell suspension was added to a volume of 1 mL per well. The cells were incubated for 4 days in a humidified atmosphere of 95% air and 5% CO_2 at 37 °C.

The in vitro viability of adherent cells cultured with the CP coating was estimated with a Countess™ Automated Cell Counter (Invitrogen, Carlsbad, CA, USA) after staining with 0.4% trypan blue (Invitrogen, USA). The percentage of viable and dead (stained) cells was measured after they were harvested with 0.05% trypsin (PanEco, Moscow, Russia) in 0.53 mM EDTA (Sigma-Aldrich, St. Louis, MO, USA) and washed twice with phosphate buffer.

2.4.2. Cytochemical Staining of HLPSCs for ALP via the Diazocoupling Technique

CP-coated titanium specimens with adherent stromal cells were air-dried, fixed for 30 s in formalin vapor, and stained with ALP. Naphthol AS-BI phosphate ($C_{18}H_{15}NO_6P$, molecular weight (m.w.) 452.21) and fast garnet GBC salt ($C_{14}H_{14}N_4O_4S$, m.w. 334.35) (both from Lachema, Brno Czech Republic) were used. Brown sites indicating enzymatic activity were considered positive cellular ALP-staining [53].

2.4.3. Immunocytochemical Detection of OCN in HLPSCs

Other CP-coated titanium plates were fixed in formalin vapor as described above. Primary antibody (rabbit polyclonal anti-human IgG (1:100), Epitomics Inc., Burlingame, CA, USA) targeting OCN and universal immunoperoxidase anti-rabbit and anti-mouse polymer (Histofine Simple Stain MAX PO MULTY, Nichirei Biosciences Inc., Tokyo, Japan) were used as described previously [52]. The appearance of brown sites in colored cells indicated positive OCN staining.

2.4.4. Computer Morphometry

A morphometry method was used to quantitatively determine cell parameters by measuring their optical features [54]. ImageJ v. 1.43 (http://www.rsb.info.nih.gov/ij) was employed to process digital images of OCN- or ALP-stained cells. Ten randomly selected images were analyzed for each sample. The dimensions of an area of stained cells were calculated (in μm^2); in addition, the areas (μm^2) of valleys and sockets that were cell-free or seeded with stained cells were measured on the CP surface.

The area of all valleys in each image was recomputed as a percent of the full image area, and the average fractional area (in %) of valleys (S_V) was calculated and is shown in for each sample with a unique average R_a.

2.5. Statistical Analysis

Correlation and regression analyses are provided; the coefficients (r) were kept at a significance level higher than 95%.

3. Results

3.1. AMSC and HLPSC Response to CP Coating

Adult (postnatal) fibroblast-like CD73CD90CD105+ AMSCs cultivated on plastic wells in a StemPro® Differentiation Kit (Thermo Fisher Scientific, USA) for 21 days were positively stained with alizarin red S (osteoblasts), alcian blue (chondrocytes), and oil red (adipocytes) (Figure 1b–d) and compared with unstained cells in standard medium (Figure 1a). The findings showed that the cells met the morphological criteria of multipotent MSCs (MMSCs) as described in [52].

The AMSCs in direct contact with microarc CP coatings in osteogenic supplement-free DMEM/F12 showed in vitro differentiation into osteoblasts (Figure 2a). Thus, a rough CP surface promotes the osteogenic potency of adult AMSCs. The osteogenic medium from the StemPro® Differentiation Kit (Thermo Fisher Scientific, USA) enhanced the osteogenic influence of the CP coating (Figure 2b; red staining). The total area of the alizarin-red-stained cells was 64,500 μm^2 for AMSCs on the CP surface in standard DMEM/F12 media (Figure 2a) and 249,500 μm^2 in osteogenic media (Figure 2b).

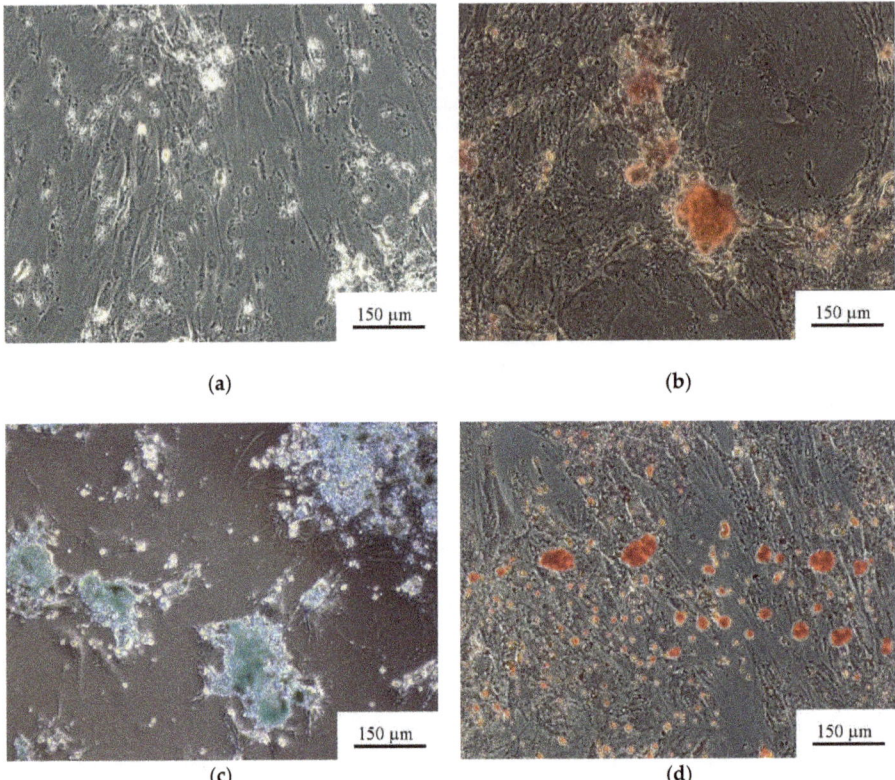

Figure 1. Human adipose-derived mesenchymal stem cells (AMSCs) cultured for 21 days in either (**a**) standard or (**b**–**d**) StemPro® differentiation media. (**a**) Some cells cultivated in the standard medium contained light fatty inclusions (unstained); (**b**) osteogenic medium, alizarin-red-stained area of mineralized intercellular regions; (**c**) chondrogenic medium, alcian-blue-stained glycoproteins; and (**d**) adipogenic medium, oil-red-stained neutral triglycerides and lipids.

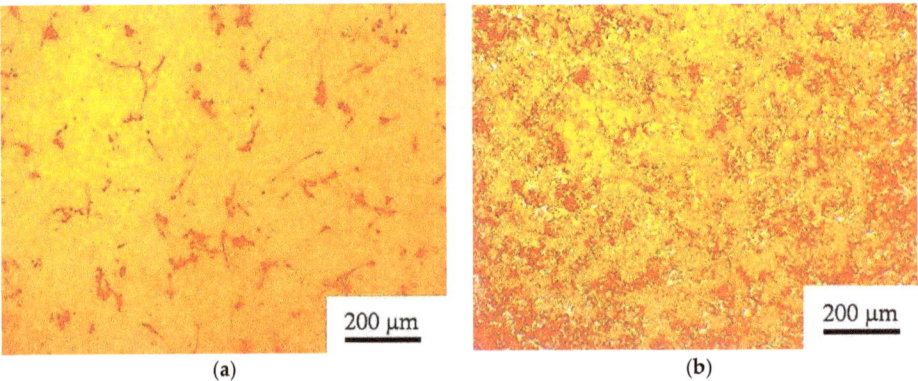

Figure 2. Human AMSCs cultured on a microarc calcium phosphate (CP) coating for 21 days in either (**a**) standard DMEM/F12 or (**b**) osteogenic differentiation media. Alizarin-red-stained areas of mineralized regions of the fibroblast-like cells and their intercellular matrix are shown.

No cytotoxicity of the CP coating was observed. Initial (before contact with samples) AMSC and HLPSC viabilities (based on the absence of trypan blue staining) were 95% and 92%. Significant differences ($p < 0.05$) in the number of viable cells were not observed for the groups studied according to a Mann–Whitney U-test (Table 1).

Table 1. Percentage of viable AMSCs or prenatal stromal cells from the human lung (HLPSCs) after cultivation with a CP-coated surface for 21 or 4 days, Me (Q_1–Q_3).

Group Studied (n = 5)	AMSC Viability, %	HLPSC Viability, %
Control cell culture	88 (88–94)	91 (88–92)
Cell culture in contact with the CP-coated sample	87 (81–90)	90 (88–91)

Note: Here, n = the number of observations (samples) in each group.

3.2. Coating Phase Composition and Morphology

XRD (Figure 3) and TEM (Figure 4) data showed that CP coatings deposited on a Ti surface by microarc oxidation are in the X-ray amorphous state, which was confirmed by the presence of a diffuse broad halo. Due to that fact, we were unable to identify the coating phase composition (Figure 3). In addition to the diffuse halo, peaks from the substrate and single CP-associated peaks were identified in the diffractograms. The crystallite size observed in the dark field images was 10–20 nm.

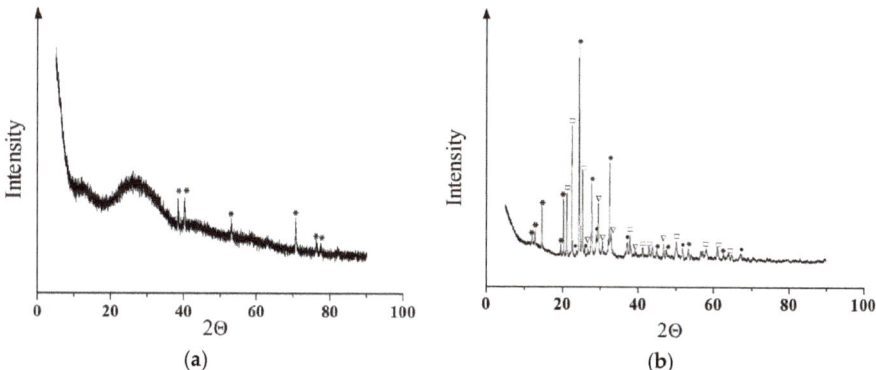

Figure 3. Diffractograms of the CP coating on a titanium surface obtained using the microarc oxidation (MAO) method. (a) The amorphous halo from the CP coating and the peaks from the titanium substrate; (b) the calcium phosphate coating after annealing at 1073 K. Peaks correspond to the phases: *—$CaTi_4(PO_4)_6$, □—TiP_2O_7, ▽—β-$Ca_2P_2O_7$, and ●—TiO_2 (anatase).

Samples were annealed in air at 1073 K for 1 h to fully identify the phase composition. The main phase of the coating was $CaTi_4(PO_4)_6$ with a fraction of β-$Ca_2P_2O_7$, TiP_2O_7 and TiO_2 (anatase) (Figures 3b and 4). Mean crystallite sizes were estimated by darkfield analysis: the mean sizes of CP phases reached 60–80 nm and those of oxides were not greater than 30–40 nm.

The microreliefs of the CP surface exhibited irregularities. The peaks of the CP microreliefs consisted of spherulites of up to 10–30 µm in diameter (Figure 5). Interconnected valleys are shown in Figure 5a as dark vast fields between ranges of bright spherulites. Single or open interconnected pores (1–10 µm in diameter) were revealed in both spherulites (Figure 5b) and valleys.

AFM measurements of the CP coating showed that its relief was embedded with 500–1000 nm submicron particles (Figure 6). The particles assembled in globules that were 1–2 µm in diameter and 30 nm in height. The positioning of the globules provided porous (pores of 1–2 µm diameter) microspherulites (3–5 µm diameter, 300 nm in height).

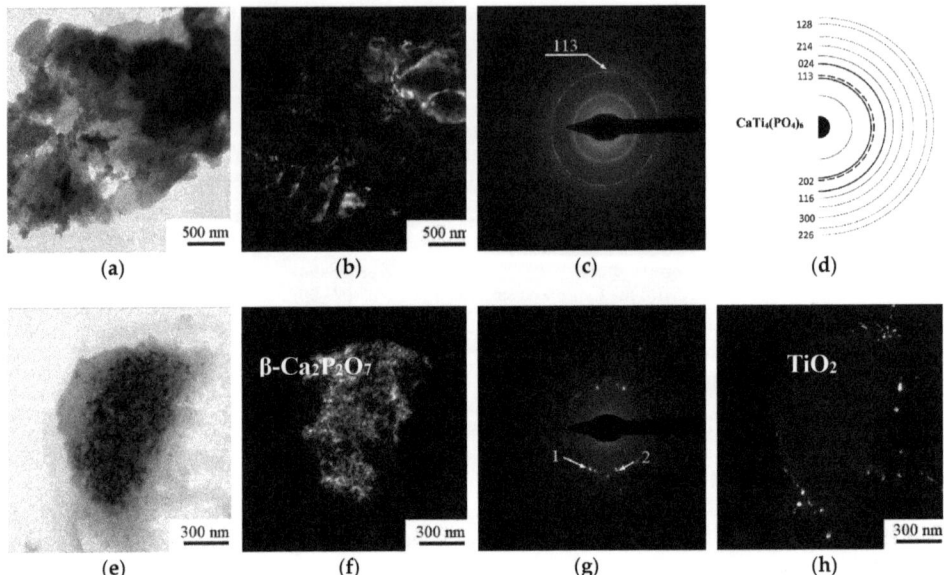

Figure 4. TEM bright field image (**a,e**) and corresponding selected area diffraction (**c,g**), dark field image (**b,f,h**), and scheme showing the interpretation (**d**) of the microarc CP coating on titanium after annealing at 1073 K.

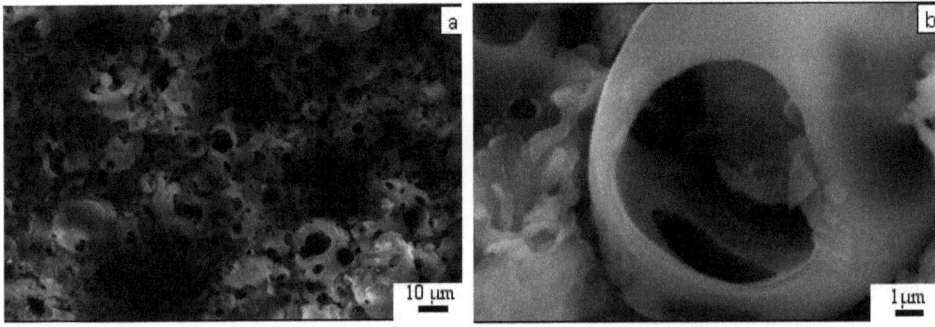

Figure 5. SEM images of the CP coating.

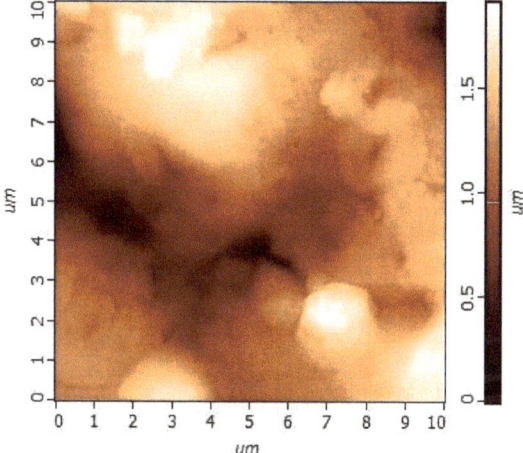

Figure 6. AFM image of the CP surface.

3.3. Relationship between Electrostatic, Geometrical, and Cytological Properties of the CP Coating

3.3.1. Microscale

V_L is correlated with the thickness (x) of the CP coatings ($r = -0.77$; significance 99%) (Figure 7). The negative value of V_L increases with increasing x. On the other hand, x is also correlated with the surface roughness index R_a ($r = 0.99$; significance >99%) (Figure 8a).

At the same time, the adhesion of the microarc-fabricated rough CP coating to the metal substrate decreased directly with increased coating thickness (Figure 8b). Thus, implants with a thick CP coating (thickness ≥40–60 µm, see Figure 8b) were biomechanically unsuccessful. Further, Figure 8c demonstrates a strong direct correlation ($r = 0.97$; significance >99%) between the amplitude of the electrode voltage (U) of the microarc device and the roughness index R_a.

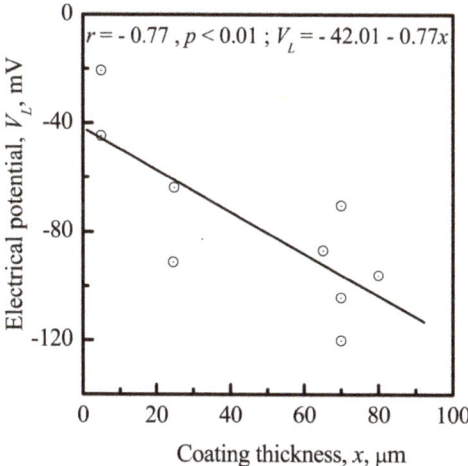

Figure 7. Correlation between the electrical potential (EP), V_L, and the CP coating thickness, x.

No correlations were observed between S_m and R_a (Figure 8d; $r = -0.06$; significance >88%) or other CP coating parameters (thickness, V_L, R_z, and R_{max}). Therefore, only the roughness index R_a (and not the index S_m) may be controlled technologically during CP coating via microarc deposition.

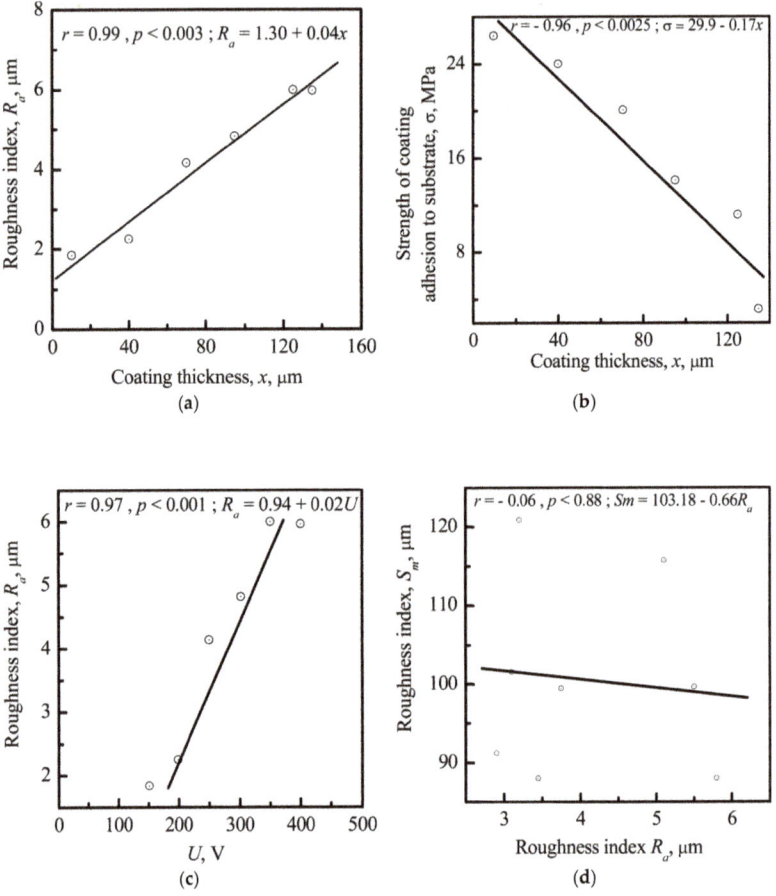

Figure 8. Correlations between (a) the CP coating thickness (x) and the roughness index R_a; (b) the CP coating thickness (x) and the adhesion strength of the CP coating to the metal substrate; (c) the amplitude of the electrode voltage (U) of the microarc device and the roughness index R_a; and (d) the CP roughness indices R_a and S_m.

Surface microtopography is biologically necessary for osteoblasts [20] to stimulate bone tissue regeneration. Previously, a microarc-fabricated rough CP coating has exhibited osteogenic activity in vivo [11,44] and in vitro [28]. Therefore, a good connection between the CP surface roughness index R_a and V_L is likely important for designing thin coatings with optimal osteogenic and biomechanical cues.

Based on the results described above, a strong correlation between V_L and R_a was sought. However, such a correlation was not found ($r = -0.33$; significance >99%), favoring the conclusion that electrical charge does not depend closely on R_a; however, a surface charge could be induced by R_{an}, i.e., at the nanoscale [45,55].

The stromal cell–surface interaction plays a dominant role in bone tissue growth. Cells can migrate into pores if the pores are more than 10 μm in diameter. However, the pores of microarc CP coatings are smaller (~1–10 μm), and MSCs likely do not seed in them. As a result, they preferentially adhere to the spherulites and valleys in the CP coating [28]. The valleys consist of single sockets (Figures 5a and 9). HLPSCs located in the sockets of the CP surface expressed the osteoblast markers ALP and OCN (Figure 9a–d). Approximately 84% of the ALP- or OCN-positive cells were found in the CP surface sockets, and only 16% were found on the spherulites.

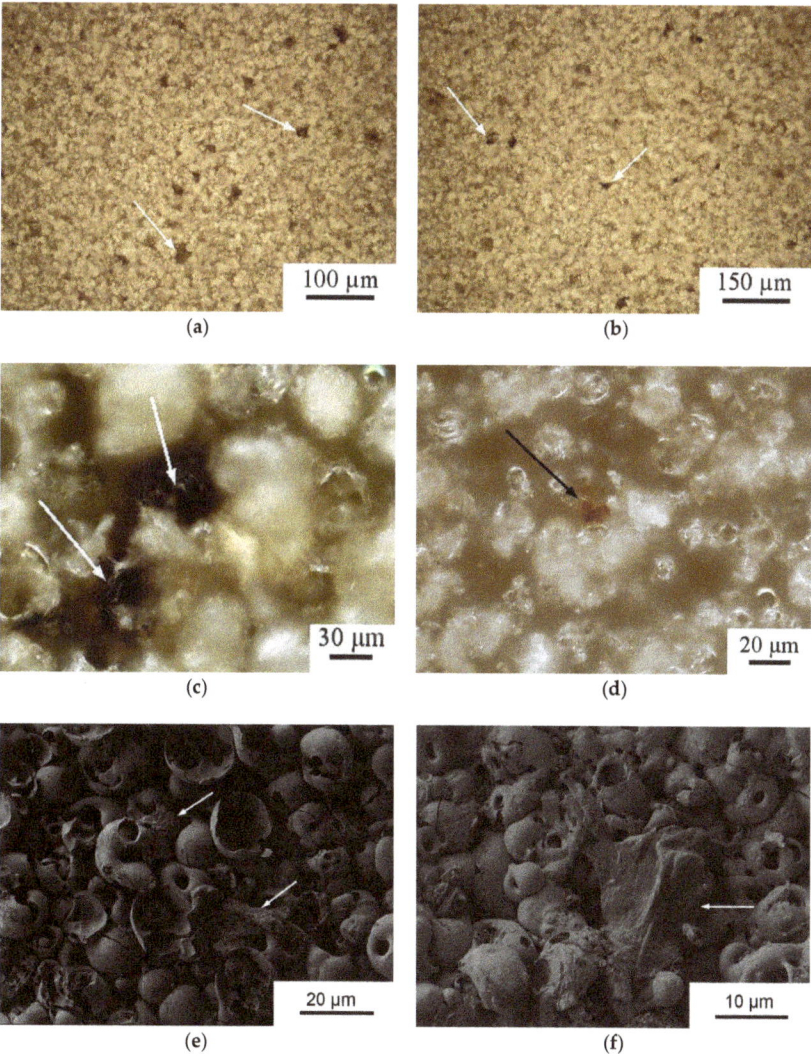

Figure 9. Surface distribution (**a**,**b**) and typical locations of alkaline phosphatase (ALP)- (**a**,**c**) or osteocalcin (OCN)-stained (**b**,**d**) HLPSCs (brown sites) in the sockets of the microarc-fabricated rough CP coating (reflecting optical microscopy) and SEM images of the HLPSCs (**e**,**f**). Cells are indicated with white and black arrows.

According to SEM, HLPSCSs are approximately 20–25 µm. EDX measurement of cell-like sites on the CP surface shows the elemental composition (78.63–79.16 C, 16.03–16.45 O, 2.84–2.94 P, 0.70–0.71 Ca, 1.26–1.28 Ti atomic %), which is unlike that of the CP coating itself (55.84 O, 25.22 P, 5.34 Ca, 13.60 Ti atomic %). Therefore, Figure 9e,f shows cells on the relief of the CP surface.

Figure 10 shows a strong correlation between the areas of ALP- ($r = 0.99$; significance >99%) or OCN-stained cells ($r = 0.91$; significance >99%) and the surrounding socket areas seeded with HLPSCs. Thus, the CP microrelief promotes an in vitro cell osteoblastic phenotype without osteogenic stimulators (β-glycerophosphate, dexamethasone, and ascorbic acid) in the culture medium.

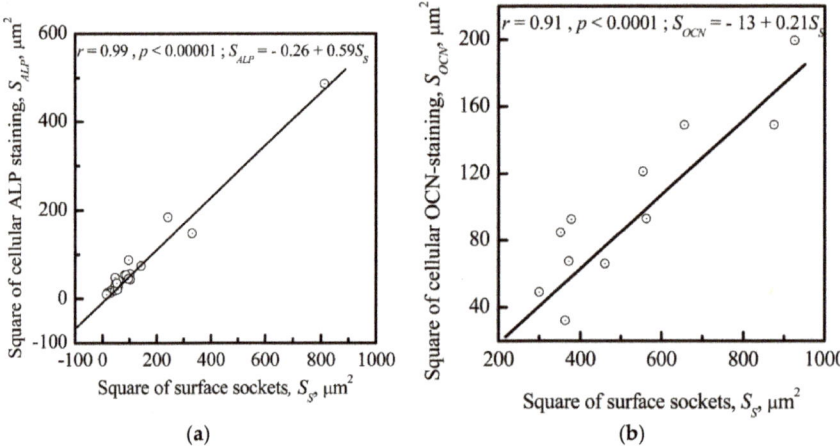

Figure 10. Correlations between the areas containing ALP-stained (a) S_{ALP} or OCN-stained (b) S_{OCN}-HLPSCs and the surrounding sockets (S_S) of the CP surface.

HLPSCs were attracted to the valleys and sockets of the microarc CP coating. The R_a of the CP surface had an effect on the S_V of the coating ($r = 0.92$; significance >99%) (Figure 11). V_L was also correlated with S_V ($r = -0.77$; significance 98%) (Figure 12).

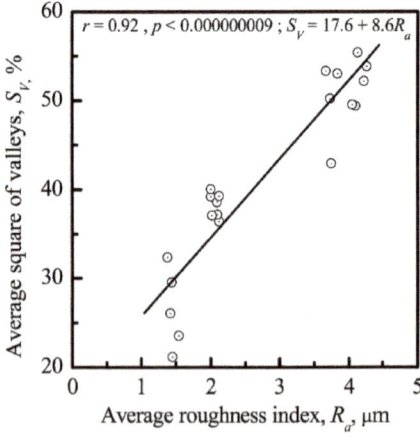

Figure 11. Correlation between the average area of valleys (S_V) and the average roughness index (R_a) of the CP surface.

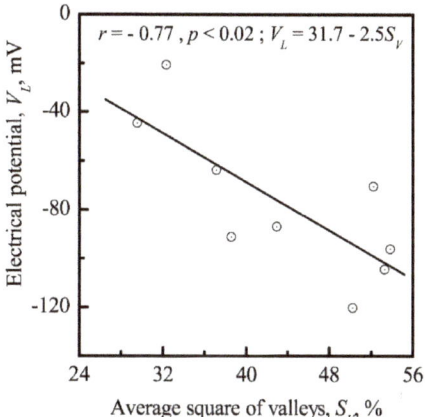

Figure 12. Correlation between the EP (V_L) and the average area of valleys (S_V) on the CP surface.

The surface valleys were formed by microsockets (Figures 5 and 9), and their areas (S_V and S_S) were correlated ($r = 0.92$; significance >99%). In addition, the correlation between V_L and S_S for rough CP surfaces was rather high ($r = -0.79$; significance >99%) (not shown).

3.3.2. The Nanoscale

The value of φ was correlated with R_{an} ($r = -0.77$; significance 95%; Figure 13), and the nanoscale potential (V_k) was correlated with φ ($r = -0.97$; significance 95%; Figure 14) for the CP surface, supporting the finding that the electrical charge density was controlled by the surface relief at the nanoscale. The sharpness of the surface topography (peaks and sockets) might be the factor that affected the electrical charge density of the CP coating.

A strong correlation between the surface electrical charge density and R_{an} (Figures 13 and 14) could explain the impact that both the electrical charge and the surface morphology of the microarc CP coating had on cell osteogenic activity.

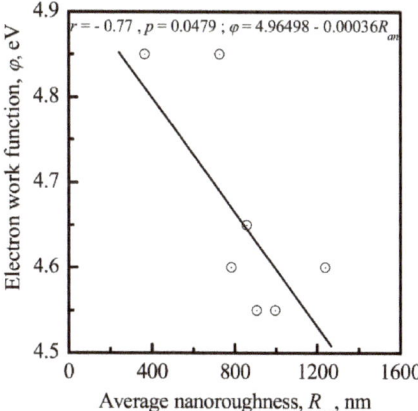

Figure 13. The dependence of the electron work function (φ) on the average nanoroughness (R_{an}) of the CP coating.

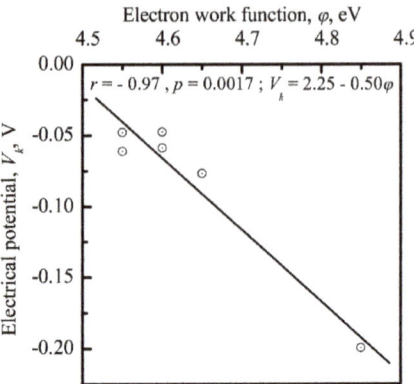

Figure 14. Relationship between the EP (V_k) and the electron work function (φ) of the CP surface at the nanoscale level.

The above microscopy results (Figure 6) demonstrate that the surface nanotopography was constructed from nanosized and submicron particles. The average size of the particles was similar for all specimens. Therefore, the deviation in CP coating irregularities was estimated as the standard deviation of R_a (SDR_a). Following the above results, a correlation of SDR_a with the standard deviation of the surface EP (SDV_L) was obtained (Figure 15). The results demonstrated a strong dependence of V_L standard deviation on the R_a standard deviation for microarc CP coating.

The sharpness of a peak can be characterized by its apex angle. The sharpness value gives an indication of the electrical field delivered by the surface electrical charges (according to general electrostatics, a higher sharpness value indicates a stronger electrical field). Figure 16 demonstrates the distribution of V_k measured by the Kelvin probe method. V_k is more positive at the peaks of the CP nanorelief than at the valleys.

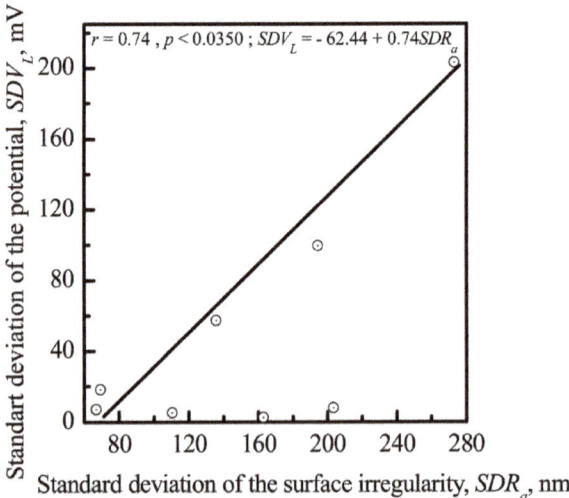

Figure 15. Correlation and interconnection of the standard deviations of surface potential (SDV_L) and surface irregularity (SDR_a) for the CP surface.

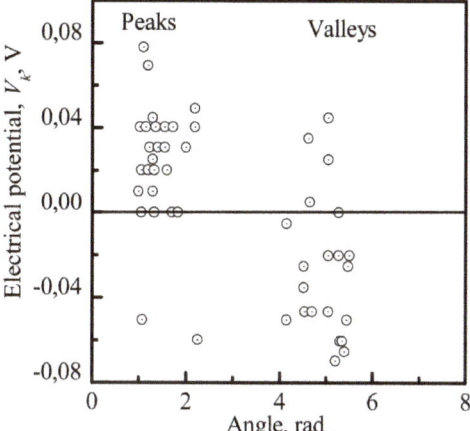

Figure 16. The distribution of the EP of the CP surface (V_k) versus the vertexes and interior angles of the CP nanorelief.

4. Discussion

Cells are immobilized in vivo within tissue and bound on a diverse array of scaffolds that are considered the extracellular matrix (ECM) of the native microenvironment [17]. The specific microenvironment where SCs exist is called the SC niche (SCN). Knowledge of the SCN and ECM control of cell fate provides tools for stimulating SC differentiation into desired cell types [29]. Biomimetic ex vivo modeling of SCNs by means of artificial materials has been attempted, and the cell behavior in such foreign ECMs has been studied [56–58]. Bone is well-known to be a native substrate for marrow MSCs, and the surface topography of the mineralized bone surface essentially affects cell fate [56]. MSCs serve as the determining component for controlling haematopoietic SCs [59].

Currently, CP surfaces have been used to imitate the mineralized bone matrix and are the most advanced ECM model. Earlier experiments have demonstrated that the microroughness of a microarc CP coating has a significant effect on the osteogenic potency of mouse bone marrow MSCs in vivo [44] and human HLPSCs in vitro [28].

MSCs are characterized as follows [52,60]: They (1) exhibit 90% viability; (2) express the surface markers CD73, CD90, and CD105 and do not express the hematopoietic markers CD45, CD34, CD20, and CD14; and (3) attach to plastic and differentiate into osteogenic, chondrogenic, and adipogenic lineages.

The in vitro culture in this work showed that adult (postnatal) fibroblast-like AMSCs positive for CD73, CD90, and CD105 met the morphological criteria for multipotent MSCs (Figure 1). In turn, a rough CP surface promoted the osteogenic potency of adult tissue-specific AMSCs (Figure 2). As a result, the occurrence of so-called "osteoblastic niches" may occur in MMSC culture [59] on CP coatings. However, AMSCs are very large cells with a size of approximately 200 µm, and their large size complicates estimation of fibroblast-like cell location and the staining area on CP relief using SEM and reflected light microscopy. Thus, HLPSCs, which are 20–25 µm, are useful for studying cell surface distribution.

Moreover, HLPSCs use in-vitro-supplied information about the early stages of bone tissue regeneration because of the ability of embryonic SCs (ESCs) to self-organize the SCN for their own development [61]. On the other hand, microenvironmental factors and, particularly, physicochemical factors regulate the self-maintenance/differentiation of ESCs in the same way as they do for tissue-specific SCs [56]. Therefore, similar effects of the nanoscale EP and the roughness of the CP coatings on SCs were studied using HLPSCs. As discussed in previous studies [22–24], an original statistical approach was developed for characterizing the relationship between surface roughness and cell adhesion. For two materials, 316L stainless steel and TiAl6V4 titanium alloy, a relationship between R_a and cell adhesion has not been found. Cell adhesion was more related to the roughness organization of substrates than to R_a. A more convenient parameter is the mean value of the profile element width, S_m. However, this correlation was insufficient, and for a more accurate characterization of cell-adhesion and roughness parameters, the new adhesion parameter "adhesion power" has been suggested. All these parameters are related in metallic materials. The situation is different for insulating materials.

The relationship between cell adhesion and roughness for CP coatings is likely more complicated because of the influences of not only surface chemistry and topography roughness organization but also of surface electrical charge, which is related to the dielectric properties of a CP coating. The size of the niche (microterritories for SCs) and its composition regulate the balance between differentiation-inducing and differentiation-inhibiting factors for both ESCs and adult SCs [61]. A quantitative assessment of the effect of CP microroughness on HLPSC differentiation into osteoblasts has led us to speculate that separate osteogenic microterritories for MSCs must exist in the ECM and may be reconstructed artificially [28]. The CP microrelief and electrical voltage have been proposed as key factors in the function of osteogenic niches [62].

However, the significance and interconnection of the CP surface properties (roughness, native charge sign, and EP amplitude) at the micro/nanoscale for human MSC osteogenic differentiation and maturation in vitro are not known.

Microscopic studies at the micro/nanorelief level have been performed on rough CP surfaces (R_a = 1.0–4.5 µm) prepared via the microarc coating technique. According to Figures 5 and 6, the microarc technology produces a three-dimensional (3D) scaffold with a bone-like thick CP coating and a multilevel structure consisting of spherulites, valleys, sockets, and pores. Thus, microarc-based simulation of the architecture of porous bone could be directed toward in vivo remodeling of the bone/bone marrow system [44,63].

Our data correspond to work on another type of CP surface [18,63–68]. CP ceramic roughness has specific osteoinductive properties [69].

Knowledge of the structure of osteogenic biomaterials that are identified by morphology and pore structure is necessary for MSC and osteoblast development and bone formation and clearly should be complemented with an understanding of the physicochemical characteristics that are extremely important to biomaterial design. There have been attempts to consider the effect of physicochemical and structural peculiarities on cellular and molecular reactions; however, the general mechanisms are not completely understood [70].

The linear relationship between the microscale surface electrical potential (V_L), thickness (x), and roughness (R_a) of the CP coating is shown in Figures 7 and 8.

The roughness spacing parameter S_m characterizes cell adhesion to the raised surface of metallic materials. In our case, no correlations were observed between the roughness index S_m and R_a (Figure 8d) or other parameters (thickness, V_L, R_z, and R_{max}) for the microarc CP coating. The roughness index R_a, unlike the S_m index, may be technically controlled during CP coating microarc deposition because of its close connection with the U of the microarc device (Figure 8c). Therefore, S_m could be used as a parameter related to cell adhesion but only for the particular case of metallic implants.

The connection of V_L and R_a to x provides a technological tool to engineer both the roughness and surface EP by controlling the CP coating thickness. This tool could be used to control cell sedimentation, migration, and adhesion to the CP surface. V_L is the factor that affects the distribution of a suspension of cells before their direct contact with a CP surface as identified by the V_L measurement obtained using the electrode lifting approach and revealed by the electrical field approximately 500 µm above the surface [50].

This process must affect the 3D features of sedimentation, distribution, and seeding and the morphofunctional state of human MSCs and HLPSCs, in particular. Different stromal cells present opposite ζ potentials, which are revealed in an external electric field. For instance, the fibroblast surface is typically negatively charged [36], while osteoblasts are characterized with a positive ζ potential [71]. Therefore, fibroblast-like cells stained with acid phosphatase and ALP-stained osteoblast-like cells could be localized at the spherulites and sockets (Figure 17), respectively, as established previously in [28].

The range in which cells are "sensitive" to the size of surface structures is usually very small and corresponds to the nano- and microscale. Detailed knowledge of their effect on cellular behavior has not been attained. A positive surface charge stimulates differentiation of osteoclast-like cells [72]. A direct electrostatic interaction between a cell and the substrate surface is considered to be a predictor of cell adhesion for implants [73]. The morphology of microarc CP coatings and its close correlation with V_L (Figures 7 and 12) could have an effect on HLPSCs, which impacts cellular colonization based on the formation of a cell–surface interface [25]. Osteoblasts respond to substrate microarchitecture [74]. Cells adhere to and proliferate on smooth surfaces (plastics, glass, and titanium) but have relatively low differentiation indices. Their proliferation decreases but differentiation increases when they are cultivated on microrough surfaces with an R_a index of 4–7 μm. Costa et al. [39] recently demonstrated the capability of biomimetic HAP coating topography to influence the attachment and differentiation of osteoblasts and the resorptive activity of osteoclasts. Osteoblast attachment and differentiation were stronger on more complex, microrough HAP surfaces (R_a ~2 μm) than on smoother topographies (R_a ~1 μm). In contrast, osteoclast activity was greater on smooth surfaces than on microrough surfaces. Thus, there are essential differences between two-dimensional (2D) and 3D cell cultures in addition to differences in cell behavior in vivo [75].

Microarc CP rough surfaces are capable of supporting an HLPSC 3D culture in the presence of osteogenic supplements (β-glycerophosphate, dexamethasone, and ascorbic acid) [28,62].

We previously named the sockets on the CP surface populated by ALP-stained (osteoblast marker) stromal cells artificial osteogenic "niches" [62]. Our culture strategy was used for the current experiments; however, osteogenic supplements were absent from the culture medium. Nevertheless, the rough CP coating alone caused HLPSC adhesion in the surface sockets and ALP and OCN expression (Figure 9). The direct dependence of ALP-staining intensity in HLPSCs on the areas of the surrounding sockets was shown (Figure 10). Thus, distinct deep microterritories of CP topography mainly promote the osteoblastic phenotype of HLPSCs.

Surface morphology determines SC attachment, spreading, proliferation, and differentiation [17,29]. However, the underlying mechanisms that trigger SC differentiation are not entirely clear [29]. AFM with a Kelvin probe was used to study the micro- and nanoscale features of the microarc CP coating.

Interconnections between the electrostatic and morphological indices of the microarc CP coating at the micro- (Figures 11 and 12) and nanoscale were discovered (Figures 13–15). The nanorelief of the microarc CP coating can define the sign and amplitude of the EP at the micro/nanoscale. Positive charges were located at the peaks of the relief, and negative polarity was observed in the valleys and sockets (Figure 16). The valleys were assembled by the microsockets (Figures 5 and 9), which led to a close correlation of the microscale V_L value with S_V of the surface valleys (Figure 12) and irregularities in the roughness (Figures 11 and 15). This correlation must affect the 3D features of human HLPSC differentiation and maturation at the micro/nanoscale.

Thus, the results revealed a nonuniform distribution of the CP topography (Figure 17a) and the charge and electric voltage values at the micro/nanoscale. We believe that the CP relief affects the surface charge sign and the magnitude of the electric voltage at the CP nano- and microsockets, and all are integrated together as joint physical factors that trigger MSC osteoblastic differentiation and maturation on the microarc CP coating (Figure 17b). The microarc coatings are X-ray amorphous and contain many primitive CP compounds ($CaTi_4(PO_4)_6$, $\beta\text{-}Ca_2P_2O_7$, TiP_2O_7, TiO_2 (anatase)) (Figures 3 and 4). One could assume that the polarity of the CP nano/microrelief is linked to the nonuniform CP phase distributions at the structural elements (spherulites and sockets) on the surface.

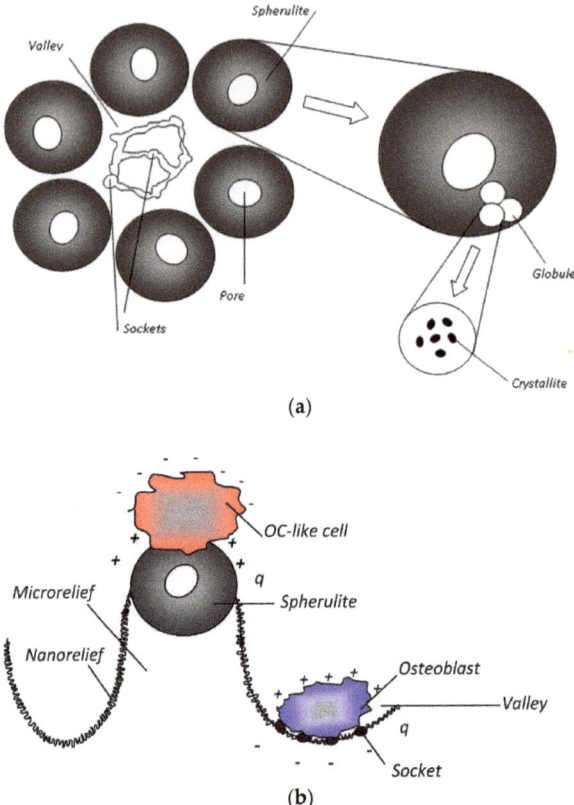

Figure 17. Schematic image of structural elements of the MAO CP coating (**a**) and nonuniform differentiation of cells in HLPSC culture (**b**). OC, osteoclast.

The electric fields play a crucial role in the fate of SCs because of their effects on cell transmembrane properties (via ion channels) and ζ potentials [30,33]. Nevertheless, the inter- and intracellular interconnections controlling MSC differentiation into osteoblasts are still not clearly understood.

Recent studies have demonstrated that multiple ion channels are heterogeneously present in the membranes of different SCs, including MSCs [33]. Therefore, hyperpolarization plays an important role in differentiation and maturation of both excitable and nonexcitable cell types. Hyperpolarization of the negative membrane potential promotes osteogenic (ALP gene expression and intracellular calcium level) differentiation of human MSCs, unlike when this membrane is depolarized [76,77]. Scaffold topography alters the intracellular calcium dynamics [78] and the state of chloride channels in cultured cells [79].

Therefore, the micro/nanoscale electric charge distribution on the microarc CP surface might affect the endogenous transmembrane potential of human MSCs through voltage-dependent ion channels and protein polarization, enabling control of cellular osteogenic differentiation and maturation. The presented topographical and electrical interconnections of the CP scaffold features are fascinating and contribute to a fundamental understanding of cell fate control in addition to being a potential tool for bone tissue engineering.

5. Conclusions

Direct interaction of cells with biomaterial surfaces is the key to surface biocompatibility [25]. Certain parameters correlate cell adhesion [80] and proliferation with the roughness of the substrate surface [10]. The present study revealed close correlations between the CP microscale surface roughness and the surface native charge, polarity, and electric potential magnitude. The negative charge and electric voltage of microarc CP coatings increase with surface thickness. Because of this phenomenon, the electric field should control the sedimentation and distribution of suspended cells before their direct contact with the CP surface.

The coating demonstrated an interconnection between CP nanoroughness and EP. The nanorelief of the microarc CP coatings defined an irregular pattern in the sign and magnitude of the EP. The nonuniform morphology of the microarc coatings led to an accumulation of negative charges within the sockets of the CP coating; however, the peaks were characterized by a positive charge. Unequal voltage distribution at the nanoscale must selectively affect cell attachment and spreading to the CP surface.

It seems that the negative charge and the magnitude of the surface EP at the nano- and microsockets ("artificial niches" [28]) are joint factors affecting the physical mechanisms that trigger MSC osteoblastic differentiation and maturation on a microarc CP coating.

The present work developed an approach involving functionalization of a substrate surface while considering both the micro/nanoscale relief ("niche-relief" concept) and EP ("niche-voltage" concept) [62] for prospective multilevel bone tissue engineering.

Author Contributions: Data curation, Y.P.S., V.V., E.V.L., K.A.Y., and K.A.P.; Formal analysis, V.V. and E.V.L.; Investigation, M.Y.K., V.V.S., O.G.K., and K.A.Y.; Project administration, Y.D. and M.Y.K.; Software, N.P.; Supervision, I.A.K. and L.S.L.; Visualization, F.T.; Writing (original draft), I.A.K.; Writing (review and editing), I.A.K., Y.D., Y.P.S., V.F.P., and L.S.L.

Funding: This research was funded by the Fundamental Research Program of the State Academies of Sciences for 2013-2020, Project no.III.23.2.5, (the MAO CP coatings portion) and by the Russian Science Foundation, Project no. 16-15-10031 (the in vitro cell culture portion).

Acknowledgments: The authors wish to thank M. Epple for fruitful discussion on the EP and surface roughness of CP coatings, Yu. Bykova for measurements of V_L, K.V. Zaitsev (Stem Cell Bank Ltd., Tomsk, Russia) for the cell line use, and Rafael Manory of www.editassociates.com for help in preparing this article for submission.

Conflicts of Interest: The authors declare no conflicts of interest.

References

1. Ratner, B.D.; Hoffman, A.S.; Schoen, F.J.; Lemons, J.E. *Biomaterials Science: An Introduction to Materials in Medicine*, 3rd ed.; Elsevier Science Publishing Co., Inc.: San Diego, CA, USA, 2012.
2. Derjaguin, B.V.; Landau, L.D. Theory of the stability of strongly charged liophobic sols and of the adhesion of strongly charged particles in solutions of electrolytes. *Acta Physicochim.* **1941**, *14*, 633–662.
3. London, F. Properties' and application of molecular forces. *Z. Phys. Chem.* **1930**, *11*, 222–251.
4. Aronov, D.; Molotskii, M.; Rosenman, G. Charge-induced wettability modification. *Appl. Phys. Lett.* **2007**, *90*, 1–3. [CrossRef]
5. Kobayashi, T.; Nakamura, S.; Yamashita, K. Enhanced osteobonding by negative surface charges of electrically polarized hydroxyapatite. *J. Biomed. Mater. Res.* **2001**, *57*, 477–484. [CrossRef]
6. Bystrov, V.S.; Coutinho, J.; Bystrova, A.V.; Dekhtyar, Y.D.; Pullar, R.C.; Poronin, A.; Palcevskis, E.; Dindune, A.; Alkan, B.; Durucan, C.; et al. Computational study of hydroxyapatite structures, properties and defects. *J. Phys. D Appl. Phys.* **2015**, *48*. [CrossRef]
7. Chen, J.Z.; Shi, Y.L.; Wang, L.; Yan, F.Y.; Zhang, F.Q. Preparation and properties of hydroxyapatite-containing titania coating by micro-arc oxidation. *Mater. Lett.* **2006**, *60*, 2538–2543. [CrossRef]
8. Frauchiger, V.M.; Schlottig, F.; Gasser, B.; Textor, M. Anodic plasma-chemical treatment of CP titanium surfaces for biomedical applications. *Biomaterials* **2004**, *25*, 593–606. [CrossRef]
9. Dawson, E.; Mapili, G.; Erickson, K.; Taqvi, S.; Roy, K. Biomaterials for stem cell differentiation. *Adv. Drug Deliv. Rev.* **2008**, *60*, 215–228. [CrossRef] [PubMed]

10. Ponsonnet, L.; Reybier, K.; Jaffrezic, N.; Comte, V.; Lagneau, C.; Lissac, M.; Martelet, C. Relationship between surface properties (roughness, wettability) of titanium and titanium alloys and cell behaviour. *Mater. Sci. Eng. C* **2003**, *23*, 551–560. [CrossRef]
11. Dekhtyar, Y.; Dvornichenko, M.V.; Karlov, A.V.; Khlusov, I.A.; Polyaka, N.; Sammons, R.; Zaytsev, K.V. Electrically functionalized hydroxyapatite and calcium phosphate surfaces to enhance immobilization and proliferation of osteoblasts in vitro and modulate osteogenesis in vivo. *IFMBE Proc.* **2009**, *25*, 245–248.
12. Martínez, E.; Engel, E.; Planell, J.A.; Samitier, J. Effects of artificial micro- and nano-structured surfaces on cell behaviour. *Ann. Anat. Anat. Anz.* **2009**, *191*, 126–135. [CrossRef] [PubMed]
13. Saha, K.; Pollock, J.F.; Schaffer, D.V.; Healy, K.E. Designing synthetic materials to control stem cell phenotype. *Curr. Opin. Chem. Biol.* **2007**, *11*, 381–387. [CrossRef] [PubMed]
14. Yu, L.M.Y.; Leipzig, N.D.; Shoichet, M.S. Promoting neuron adhesion and growth. *Mater. Today* **2008**, *11*, 36–43. [CrossRef]
15. Palcevskis, E.; Dindune, A.; Dekhtyar, Y.; Polyaka, N.; Veljović, D.; Sammons, R.L. The influence of surface treatment by hydrogenation on the biocompatibility of different hydroxyapatite materials. *IOP Conf. Ser. Mater. Sci. Eng.* **2011**, *23*, 012032. [CrossRef]
16. Curtis, A.; Varde, M. Control of Cell Behavior: Topological Factors. *J. Natl. Cancer Inst.* **1964**, *33*, 15–62. [PubMed]
17. Sniadecki, N.J.; Desai, R.A.; Ruiz, S.A.; Chen, C.S. Nanotechnology for cell-substrate interactions. *Ann. Biomed. Eng.* **2006**, *34*, 59–74. [CrossRef] [PubMed]
18. Klein, C.; de Groot, K.; Chen, W.; Li, Y.; Zhang, X. Osseous substance formation induced in porous calcium phosphate ceramics in soft tissues. *Biomaterials* **1994**, *15*, 31–34. [CrossRef]
19. Anselme, K. Osteoblast adhesion on biomaterials. *Biomaterials* **2000**, *21*, 667–681. [CrossRef]
20. Anselme, K.; Bigerelle, M.; Noel, B.; Dufresne, E.; Judas, D.; Iost, A.; Hardouin, P. Qualitative and quantitative study of human osteoblast adhesion on materials with various surface roughnesses. *J. Biomed. Mater. Res.* **2000**, *49*, 155–166. [CrossRef]
21. Anselme, K.; Bigerelle, M. Topography effects of pure titanium substrates on human osteoblast long-term adhesion. *Acta Biomater.* **2005**, *1*, 211–222. [CrossRef] [PubMed]
22. Giljean, S.; Ponche, A.; Bigerelle, M.; Anselme, K. Statistical approach of chemistry and topography effect on human osteoblast adhesion. *J. Biomed. Mater. Res.-A* **2010**, *94*, 1111–1123. [CrossRef] [PubMed]
23. Anselme, K.; Bigerelle, M. On the relation between surface roughness of metallic substrates and adhesion of human primary bone cells. *Scanning* **2014**, *36*, 11–20. [CrossRef] [PubMed]
24. Giljean, S.; Bigerelle, M.; Anselme, K. Roughness statistical influence on cell adhesion using profilometry and multiscale analysis. *Scanning* **2014**, *36*, 2–10. [CrossRef] [PubMed]
25. Curtis, A.; Wilkinson, C. Topographical control of cells. *Biomaterials* **1997**, *18*, 1573–1583. [CrossRef]
26. Schwartz, Z.; Nasazky, E.; Boyan, B.D. Surface microtopography regulates osteointegration: The role of implant surface microtopography in osteointegration. *Alpha Omegan* **2005**, *98*, 9–19. [PubMed]
27. Meyer, U.; Büchter, A.; Wiesmann, H.P.; Joos, U.; Jones, D.B. Basic reactions of osteoblasts on structured material surfaces. *Eur. Cells Mater.* **2005**, *9*, 39–49. [CrossRef]
28. Khlusov, I.A.; Khlusova, M.Y.; Zaitsev, K.V.; Kolokol'tsova, T.D.; Sharkeev, Y.P.; Pichugin, V.F.; Legostaeva, E.V.; Trofimova, I.E.; Klimov, A.S.; Zhdanova, A.I. Pilot in vitro study of the parameters of artificial niche for osteogenic differentiation of human stromal stem cell pool. *Bull. Exp. Biol. Med.* **2011**, *150*, 535–542. [CrossRef] [PubMed]
29. Ravichandran, R. Effects of nanotopography on stem cell phenotypes. *World J. Stem Cells* **2009**, *1*, 55. [CrossRef] [PubMed]
30. Heubach, J.F.; Graf, E.M.; Leutheuser, J.; Bock, M.; Balana, B.; Zahanich, I.; Christ, T.; Boxberger, S.; Wettwer, E.; Ravens, U. Electrophysiological properties of human mesenchymal stem cells. *J. Physiol.* **2004**, *554*, 659–672. [CrossRef] [PubMed]
31. McCaig, C.D.; Rajnicek, A.M.; Song, B.; Zhao, M. Controlling cell behavior electrically: Current views and future potential. *Physiol. Rev.* **2005**, *85*, 943–978. [CrossRef] [PubMed]
32. Levin, M. Large-scale biophysics: Ion flows and regeneration. *Trends Cell Biol.* **2007**, *17*, 261–270. [CrossRef] [PubMed]
33. Li, G.-R. Functional ion channels in stem cells. *World J. Stem Cells* **2011**, *3*, 19. [CrossRef] [PubMed]

34. Cai, K.; Frant, M.; Bossert, J.; Hildebrand, G.; Liefeith, K.; Jandt, K.D. Surface functionalized titanium thin films: Zeta-potential, protein adsorption and cell proliferation. *Colloids Surf. B Biointerfaces* **2006**, *50*, 1–8. [CrossRef] [PubMed]
35. Chung, T.H.; Wu, S.H.; Yao, M.; Lu, C.W.; Lin, Y.S.; Hung, Y.; Mou, C.Y.; Chen, Y.C.; Huang, D.M. The effect of surface charge on the uptake and biological function of mesoporous silica nanoparticles in 3T3-L1 cells and human mesenchymal stem cells. *Biomaterials* **2007**, *28*, 2959–2966. [CrossRef] [PubMed]
36. Hamdan, M.; Blanco, L.; Khraisat, A.; Tresguerres, I.F. Influence of Titanium Surface Charge on Fibroblast Adhesion. *Clin. Implant Dent. Relat. Res.* **2006**, *8*, 32–38. [CrossRef] [PubMed]
37. Nakamura, M.; Nagai, A.; Hentunen, T.; Salonen, J.; Sekijima, Y.; Okura, T.; Hashimoto, K.; Toda, Y.; Monma, H.; Yamashita, K. Surface electric fields increase osteoblast adhesion through improved wettability on hydroxyapatite electret. *ACS Appl. Mater. Interfaces* **2009**, *1*, 2181–2189. [CrossRef] [PubMed]
38. Thian, E.S.; Ahmad, Z.; Huang, J.; Edirisinghe, M.J.; Jayasinghe, S.N.; Ireland, D.C.; Brooks, R.A.; Rushton, N.; Bonfield, W.; Best, S.M. Electrosprayed nanoapatite: A new generation of bioactive material. *Key Eng. Mater.* **2008**, *361–363*, 597. [CrossRef]
39. Costa, D.O.; Prowse, P.D.H.; Chrones, T.; Sims, S.M.; Hamilton, D.W.; Rizkalla, A.S.; Dixon, S.J. The differential regulation of osteoblast and osteoclast activity bysurface topography of hydroxyapatite coatings. *Biomaterials* **2013**, *34*, 7215–7226. [CrossRef] [PubMed]
40. Cai, K.; Bossert, J.; Jandt, K.D. Does the nanometre scale topography of titanium influence protein adsorption and cell proliferation? *Colloids Surf. B Biointerfaces* **2006**, *49*, 136–144. [CrossRef] [PubMed]
41. Huang, D.-M.; Chung, T.-H.; Hung, Y.; Lu, F.; Wu, S.-H.; Mou, C.-Y.; Yao, M.; Chen, Y.-C. Internalization of mesoporous silica nanoparticles induces transient but not sufficient osteogenic signals in human mesenchymal stem cells. *Toxicol. Appl. Pharmacol.* **2008**, *231*, 208–215. [CrossRef] [PubMed]
42. Aronov, D.; Rosenman, G.; Karlov, A.; Shashkin, A. Wettability patterning of hydroxyapatite nanobioceramics induced by surface potential modification. *Appl. Phys. Lett.* **2006**, *88*. [CrossRef]
43. Aronov, D.; Rosen, R.; Ron, E.Z.; Rosenman, G. Tunable hydroxyapatite wettability: Effect on adhesion of biological molecules. *Process Biochem.* **2006**, *41*, 2367–2372. [CrossRef]
44. Khlusov, I.A.; Karlov, A.V.; Sharkeev, Y.P.; Pichugin, V.F.; Kolobov, Y.P.; Shashkina, G.A.; Ivanov, M.B.; Legostaeva, E.V.; Sukhikh, G.T. Osteogenic potential of mesenchymal stem cells from bone marrow in situ: Role of physicochemical properties of artificial surfaces. *Bull. Exp. Biol. Med.* **2005**, *140*, 144–152. [CrossRef] [PubMed]
45. Elter, P.; Thull, R. Protein adsorption on nanostructured implant surfaces: A model calculation for the prediction of preferred adsorption sites. *Biomaterials* **2006**, *7*, 138–144.
46. Sharkeev, Y.P.; Legostaeva, E.V.; Eroshenko, Y.A.; Khlusov, I.A.; Kashin, O.A. The Structure and Physical and Mechanical Properties of a Novel Biocomposite Material, Nanostructured Titanium–Calcium-Phosphate Coating. *Compos. Interfaces* **2009**, *16*, 535–546. [CrossRef]
47. Legostaeva, E.V.; Kulyashova, K.S.; Komarova, E.G.; Epple, M.; Sharkeev, Y.P.; Khlusov, I.A. Physical, chemical and biological properties of micro-arc deposited calcium phosphate coatings on titanium and zirconium-niobium alloy. *Materwiss. Werksttech.* **2013**, *44*, 188–197. [CrossRef]
48. Leach, R.K. *The Measurement of Surface Texture Using Stylus Instruments*; National Physical Laboratory Teddington: Middlesex, UK, 2001; p. 97.
49. Eguchi, M. On the permanent electret. *Philos. Mag.* **1925**, *49*, 178–192. [CrossRef]
50. Gostischev, E.A. A Noncontacting Manner to Determine the Potentials of Charged Surface of Objects and the Device for Its Realization. Russian Federation Patent 2223511, 10 February 2004.
51. Akmene, R.J.; Balodis, A.J.; Dekhtyar, Y.D.; Markelova, G.N.; Matvejevs, J.V.; Rozenfelds, L.B.; Sagalovias, G.L.; Smirnovs, J.S.; Tolkaaovs, A.A.; Upmiņš, A.I. Exoelectron emission spectrometre complete set of surface local investigation. *Poverhn. Fiz. Him. Meh.* **1993**, *8*, 125–128. (In Russian)
52. Bourin, P.; Bunnell, B.A.; Casteilla, L.; Dominici, M.; Katz, A.J.; March, K.L.; Redl, H.; Rubin, J.P.; Yoshimura, K.; Gimble, J.M. Stromal cells from the adipose tissue-derived stromal vascular fraction and culture expanded adipose tissue-derived stromal/stem cells: A joint statement of the International Federation for Adipose Therapeutics and Science (IFATS) and the International Society for Cellular Therapy (ISCT). *Cytotherapy* **2013**, *15*, 641–648. [CrossRef] [PubMed]

53. Khlusov, I.A.; Shevtsova, N.M.; Khlusova, M.Y. Detection in vitro and quantitative estimation of artificial microterritories which promote osteogenic differentiation and maturation of stromal stem cells. *Methods Mol. Biol.* **2013**, *1035*, 103–119. [CrossRef] [PubMed]
54. Freshney, R.I. *Culture of Animal Cells: A Manual of Basic Technique and Specialized Applications*, 6th ed.; John Wiley & Sons, Inc.: Hoboken, NJ, USA, 2011; ISBN 9780470528129.
55. Bystrov, V.S.; Paramonova, E.; Dekhtyar, Y.; Katashev, A.; Karlov, A.; Polyaka, N.; Bystrova, A.V.; Patmalnieks, A.; Kholkin, A.L. Computational and experimental studies of size and shape related physical properties of hydroxyapatite nanoparticles. *J. Phys. Condens. Matter* **2011**, *23*, 65302. [CrossRef] [PubMed]
56. Dellatore, S.M.; Garcia, A.S.; Miller, W.M. Mimicking stem cell niches to increase stem cell expansion. *Curr. Opin. Biotechnol.* **2008**, *19*, 534–540. [CrossRef] [PubMed]
57. Lutolf, M.P.; Gilbert, P.M.; Blau, H.M. Designing materials to direct stem-cell fate. *Nature* **2009**, *462*, 433–441. [CrossRef] [PubMed]
58. Lutolf, M.P.; Doyonnas, R.; Havenstrite, K.; Koleckar, K.; Blau, H.M. Perturbation of single hematopoietic stem cell fates in artificial niches. *Integr. Biol.* **2009**, *1*, 59–69. [CrossRef] [PubMed]
59. Jing, D.; Fonseca, A.V.; Alakel, N.; Fierro, F.A.; Muller, K.; Bornhauser, M.; Ehninger, G.; Corbeil, D.; Ordemann, R. Hematopoietic stem cells in co-culture with mesenchymal stromal cells–modeling the niche compartments in vitro. *Haematologica* **2010**, *95*, 542–550. [CrossRef] [PubMed]
60. Dominici, M.; Le Blanc, K.; Mueller, I.; Slaper-Cortenbach, I.; Marini, F.C.; Krause, D.S.; Deans, R.J.; Keating, A.; Prockop, D.J.; Horwitz, E.M. Minimal criteria for defining multipotent mesenchymal stromal cells. The International Society for Cellular Therapy position statement. *Cytotherapy* **2006**, *8*, 315–317. [CrossRef] [PubMed]
61. Peerani, R.; Rao, B.M.; Bauwens, C.; Yin, T.; Wood, G.A.; Nagy, A.; Kumacheva, E.; Zandstra, P.W. Niche-mediated control of human embryonic stem cell self-renewal and differentiation. *EMBO J.* **2007**, *26*, 4744–4755. [CrossRef] [PubMed]
62. Khlusov, I.A.; Dekhtyar, Y.; Khlusova, M.Y.; Gostischev, E.A.; Sharkeev, Y.P.; Pichugin, V.F.; Legostaeva, E.V. Novel concepts of "niche-relief" and "niche-voltage" for stem cells as a base of bone and hematopoietic tissues biomimetic engineering. *IFMBE Proc.* **2013**, *38*, 99–102.
63. Terleeva, O.P.; Sharkeev, Y.P.; Slonova, A.I.; Mironov, I.V.; Legostaeva, E.V.; Khlusov, I.A.; Matykina, E.; Skeldon, P.; Thompson, G.E. Effect of microplasma modes and electrolyte composition on micro-arc oxidation coatings on titanium for medical applications. *Surf. Coat. Technol.* **2010**, *205*, 1723–1729. [CrossRef]
64. De Groot, K. Bioceramics consisting of calcium phosphate salts. *Biomaterials* **1980**, *1*, 47–50. [CrossRef]
65. De Bruijn, J.D. *Calcium Phosphate Biomaterials: Bone-Bonding and Biodegradation Properties*; Leiden University: Leiden, The Netherlands, 1993.
66. Ikeda, N.; Kawanabe, K.; Nakamura, T. Quantitative comparison of osteoconduction of porous, dense A-W glass-ceramic and hydroxyapatite granules (effects of granule and pore sizes). *Biomaterials* **1999**, *20*, 1087–1095. [CrossRef]
67. Gauthier, O.; Bouler, J.M.; Aguado, E.; Pilet, P.; Daculsi, G. Macroporous biphasic calcium phosphate ceramics: Influence of macropore diameter and macroporosity percentage on bone ingrowth. *Biomaterials* **1998**, *19*, 133–139. [CrossRef]
68. Sous, M. Cellular biocompatibility and resistance to compression of macroporous β-tricalcium phosphate ceramics. *Biomaterials* **1998**, *19*, 2147–2153. [CrossRef]
69. Yuan, H.; Kurashina, K.; De Bruijn, J.D.; Li, Y.; De Groot, K.; Zhang, X. A preliminary study on osteoinduction of two kinds of calcium phosphate ceramics. *Biomaterials* **1999**, *20*, 1799–1806. [CrossRef]
70. Lobo, S.E.; Arinzeh, T.L. Biphasic calcium phosphate ceramics for bone regeneration and tissue engineering applications. *Materials (Basel)* **2010**, *3*, 815–826. [CrossRef]
71. Ferrier, J.; Ross, S.M.; Kanehisa, J.; Aubin, J.E. Osteoclasts and osteoblasts migrate in opposite directions in response to a constant electrical field. *J. Cell. Physiol.* **1986**, *129*, 283–288. [CrossRef] [PubMed]
72. Hamamoto, N.; Hamamoto, Y.; Nakajima, T.; Ozawa, H. Histological, histochemical and ultrastructural study on the effects of surface charge on bone formation in the rabbit mandible. *Arch. Oral Biol.* **1995**, *40*, 97–106. [CrossRef]
73. Smith, I.O.; Baumann, M.J.; McCabe, L.R. Electrostatic interactions as a predictor for osteoblast attachment to biomaterials. *J. Biomed. Mater. Res. A* **2004**, *70*, 436–441. [CrossRef] [PubMed]

74. Boyan, B.D.; Lossdorfer, S.; Wang, L.; Zhao, G.; Lohmann, C.H.; Cochran, D.L.; Schwartz, Z. Osteoblasts generate an osteogenic microenvironment when grown on surfaces with rough microtopographies. *Eur. Cells Mater.* **2003**, *6*, 22–27. [CrossRef]
75. Birgersdotter, A.; Sandberg, R.; Ernberg, I. Gene expression perturbation in vitro—A growing case for three-dimensional (3D) culture systems. *Semin. Cancer Biol.* **2005**, *15*, 405–412. [CrossRef] [PubMed]
76. Sundelacruz, S.; Levin, M.; Kaplan, D.L. Membrane potential controls adipogenic and osteogenic differentiation of mesenchymal stem cells. *PLoS ONE* **2008**, *3*. [CrossRef] [PubMed]
77. Sundelacruz, S.; Levin, M.; Kaplan, D.L. Role of membrane potential in the regulation of cell proliferation and differentiation. *Stem Cell Rev. Rep.* **2009**, *5*, 231–246. [CrossRef] [PubMed]
78. Yin, L. Scaffold topography alters intracellular calcium dynamics in cultured cardiomyocyte networks. *AJP Heart Circ. Physiol.* **2004**, *287*, H1276–H1285. [CrossRef] [PubMed]
79. Tobasnick, G.; Curtis, A. Chloride channels and cell topographic reaction. *Eur. Cells Mater.* **2001**, *2*, 49–61. [CrossRef]
80. Hallab, N.J.; Bundy, K.J.; O'Connor, K.; Clark, R.; Moses, R.L. Cell adhesion to biomaterials: Correlations between surface charge, surface roughness, adsorbed protein, and cell morphology. *J. Long-Term Eff. Med. Implant.* **1995**, *5*, 209–231.

© 2018 by the authors. Licensee MDPI, Basel, Switzerland. This article is an open access article distributed under the terms and conditions of the Creative Commons Attribution (CC BY) license (http://creativecommons.org/licenses/by/4.0/).

Article

Rheological and Mechanical Properties of Thermoresponsive Methylcellulose/Calcium Phosphate-Based Injectable Bone Substitutes

Öznur Demir Oğuz and Duygu Ege *

Institute of Biomedical Engineering, Boğaziçi University, Rasathane St., Kandilli, 34684 İstanbul, Turkey; oznur.demir@boun.edu.tr
* Correspondence: duygu.ege@boun.edu.tr; Tel.: +90-216-516-3438

Received: 26 February 2018; Accepted: 27 March 2018; Published: 14 April 2018

Abstract: In this study, a novel injectable bone substitute (IBS) was prepared by incorporating a bioceramic powder in a polymeric solution comprising of methylcellulose (MC), gelatin and citric acid. Methylcellulose was utilized as the polymeric matrix due to its thermoresponsive properties and biocompatibility. 2.5 wt % gelatin and 3 wt % citric acid were added to the MC to adjust the rheological properties of the prepared IBS. Then, 0, 20, 30 and 50 wt % of the bioceramic component comprising tetracalcium phosphate/hydroxyapatite (TTCP/HA), dicalcium phosphate dehydrate (DCPD) and calcium sulfate dehydrate (CSD) were added into the prepared polymeric component. The prepared IBS samples had a chewing gum like consistency. IBS samples were investigated in terms of their chemical structure, rheological characteristics, and mechanical properties. After that, in vitro degradation studies were carried out by measurement of pH and % remaining weight. Viscoelastic characteristics of the samples indicated that all of the prepared IBS were injectable and they hardened at approximately 37 °C. Moreover, with increasing wt % of the bioceramic component, the degradation rate of the samples significantly reduced and the mechanical properties were improved. Therefore, the experimental results indicated that the P50 mix may be a promising candidates to fill bone defects and assist bone recovery for non-load bearing applications.

Keywords: methylcellulose; injectable bone substitutes; calcium phosphate cement; calcium sulfate; citric acid; gelatin; bone; rheological studies; injectability; mechanical properties

1. Introduction

Since the late 1980s, there is ongoing research on the development of injectable bone substitutes (IBS) [1–3]. The most commonly studied IBS's in dentistry and orthopedic applications are calcium phosphate cements (CPC) due to their excellent physical, mechanical and biological properties [3–5]. Since CPCs are biocompatible, bioactive, biodegradable and osteoconductive, there is tremendous research on their further development. CPCs are injectable, and they harden in vivo, after taking the shape of the defect site. Two of the most commonly used bioceramics for the production of CPCs are dicalcium phosphate dihydrate (DCPD, $CaHPO_4.2H_2O$) and tetra-calcium phosphate (TTCP, $Ca_4(PO_4)_2O$) [4,6,7]. After a mixture of TTCP and DCPD is injected into the defect site, it transforms into hydroxyapatite (HA) [6–9]. Pure CPC bone substitutes have many drawbacks; such as the possibility of collapse under physiological conditions, poor degradability, as well as weak torsional and bending strength [10–12]. The addition of calcium sulfates (CSD) to CPCs may overcome some of these drawbacks, including the improvement of their injectability, mechanical strength, degradability, and osteointegration. CSD form after mixing α-calcium sulfate hemihydrate with distilled water. The addition of CSD into CPC leads to improvement of plasticity and the compression strength of the CPC [11,13,14].

To adjust the rheological properties, including injectability, setting temperature, mechanical properties of IBS, CPC and CSD mixture may be incorporated into a polymeric matrix. The presence of a polymeric matrix may also assist the permeation of body fluids into bone substitutes; therefore promoting three-dimensional cell migration, cell growth and ultimately ossification [4,15]. Cellulose is a promising polymer for this purpose; however, it is not water-soluble due to the presence of intra-molecular hydrogen bonding; which restricts its range of biomedical applications [16–18]. Therefore, hydrophilic and water-soluble derivatives of cellulose have emerged [19]. Methylcellulose (MC), which is one of the cellulose ether derivatives, has been extensively studied for biomedical applications. It also shows the potential for production of IBS due to its thermoresponsive properties [16]. Research shows that MC enhances injectability, mechanical properties, and osteoconductivity of CPC-based IBS [20–23]. In their study, Liu et al. used 8, 10 and 12 wt % of MC to prepare injectable, biocompatible and biodegradable hydrogels. Overall, 10 wt % MC led to the most suitable rheological properties for biomedical applications including viscosity, setting time and storage modulus [24].

In addition to that, like MC, gelatin has been added to CPC pastes; which does not only mimics the organic matrix of bone, but also contributes to the hardening of CPC and improves the workability of cement pastes [25,26]. Moreover, the addition of gelatin into MC enables gelation at the physiological temperature and also provides improved cell-surface interactions due to the presence of its arginine-glycine-aspartate or RGD group [26–28]. Nishinari et al. [29] observed that the interaction between the non-substituted hydroxyl group in methylcellulose and the carboxyl group in gelatin reduces the gelling temperature of MC. [9,30]. In many studies, 2–3 wt % gelatin was added in hydrogels to improve their biocompatibility. Additionally, gelatin addition improved the mechanical properties of hydrogels [3,31,32].

Experimental studies show that the addition of different salts to MC solution also reduces the gelling temperature to the physiological temperature [16,19,33]. Citric acid is a cost-effective, non-toxic additive which optimizes the physical properties of IBS; such as degree of swelling, setting time, viscosity and mechanical properties [9,30]. The carboxylate group in citric acid reacts with the hydroxyl groups of MC and the amino groups of gelatin via esterification. Citric acid acts as a liquefier and therefore also enhances the injectability and workability of CPC pastes [34,35]. It also improves the strength of CPC pastes due to its high salting-out effect according to the Hofmeister series [9]. In previous studies, citric acid was incorporated in calcium phosphate cements up to 3 wt%. With an increase of wt % of citric acid, setting time and injectability decreased which was described as acceptable for clinical applications [36].

In the current study, IBS was prepared by mixing MC, gelatin, citric acid, and different wt % of bioceramic comprising of CPC and CSD. As previous studies indicated improved physical properties for use of 10 wt % MC, 3 wt % citric acid and 2.5 wt % gelatin in prepared scaffolds; in this study, these values were kept constant. With this novel combination of polymers and CPC, an ideal IBS for orthopedic applications can be developed which stimulates new bone formation. 20, 30, and 50 wt % of CPC/CSD–based bioceramic powder was added in the polymeric matrix to produce IBS. The chemical structure of the samples was investigated using X-ray Diffraction (XRD). The physical handling and mechanical properties of the IBS samples were analyzed with rheological studies, compressive strength measurements and injectability tests. After that, the in vitro degradation studies were carried out. Overall, the experimental studies suggest that MC/bioceramic powder-based IBS were promising in terms of their rheological, mechanical and degradation properties for future bone treatment applications.

2. Materials and Methods

2.1. Materials

Methylcellulose (MC, viscosity 15 cps, Mw 14 kDa, DS 1.5–1.9), gelatin (from bovine skin, gel strength ~225 g Bloom, Type B), sodium citrate tribasic dihydrate (SC, $C_6H_5Na_3O_7$ Mw 294.10 g/mol, mp: >300 °C (lit.), pH 7.0–9.0 at 25 °C), calcium hydrogen phosphate dihydrate (DCPD, $CaHPO_4 \cdot 2H_2O$, Mw 172.09 g/mol, d:2.31 g/mL), monetite ($CaHPO_4$, Mw 136.06 g/mol), calcium carbonate ($CaCO_3$, Mw 100.09 g/mol, d: 2.93 g, ≤30 µm particle size) and phosphate buffer saline (PBS, tablets, pH 7.4 at 25 °C) were purchased from Sigma Aldrich (Sigma Aldrich, Taufkirchen, Germany). Calcium sulfate dihydrate (CSD, $CaSO_4 \cdot 2H_2O$, Mw 172.17 g/mol) was purchased from Merck (Merck KGaA, Darmstadt, Germany).

2.2. Preparation of Injectable Bone Substitutes (IBS) Samples

2.2.1. Preparation of the Polymeric Solution

Methylcellulose (MC) solution was prepared by dissolving 6 g of MC powder in distilled water at 90 °C until MC's complete dissolution. The prepared solution was stored at 4 °C overnight to obtain a clear MC solution [33]. 1.875 g of gelatin was dissolved in 12.5 mL of distilled water at 50 °C and was allowed to cool down before use [24]. 2.25 g of sodium citrate dihydrates (SC) solution was prepared by dissolving SC salts in 12.5 mL distilled water at room temperature [19]. Then all solutions were blended to a final concentration of 8.0, 2.5, and 3.0 wt % of MC, gelatin, and SC, respectively.

2.2.2. Preparation of the Bioceramic Powder Mixture

The bioceramic component of IBS samples consist of TTCP/HA-based powder, DCPD, and CSD. Analytical grade DCPD and CSD were used without any further purification. TTCP/HA-based powder was synthesized by heating an equi-molar mixture of monetite ($CaHPO_4$) and calcium carbonate ($CaCO_3$) at 1500 °C for 6 h, using a 5 °C/min heating rate and 10 °C/min cooling rate [37]. CPC mixture was prepared by mixing an equi-molar mixture of DCPD and TTCP/HA-based powder and then this CPC mixture was added to the CSD in a 4 to 1 mass ratio [38].

After this, different wt % of the bioceramic powder phase was added to the polymeric component. Table 1 shows the compositions of prepared IBS samples.

Table 1. Composition of the injectable bone substitutes (IBS) samples.

Abbreviation	Gelatin (wt %)	SC (wt %)	MC (wt %)	Bioceramic Powder Component (wt %)
P0	2.5	3	8	0
P20	2.5	3	8	20
P30	2.5	3	8	30
P50	2.5	3	8	50

2.3. Characterization of IBS Samples

2.3.1. XRD Analysis

XRD (Rigaku D/MAX-Ultima+/PC, Austin, TX, USA) analysis was performed to examine the synthesized powder and lyophilized IBS samples at 40 kV and 30 mA with a step size of 0.01° between 20° and 40° in a fixed time mode at Bogaziçi University, Istanbul, Turkey [33].

2.3.2. Fourier Transform Infrared Spectroscopy (FTIR) Analysis

Functional groups of the synthesized powder were detected by using a Perkin Elmer FTIR spectrometer (Perkin Elmer, Waltham, MA, USA) from 4000 to 400 cm^{-1} using the KBr pellet technique at Yildiz Technical University, Istanbul, Turkey [39].

2.3.3. Injectability Measurements

Injectability of the samples was qualitatively evaluated by extruding the IBS samples through a disposable syringe, using an 18-gauge needle, in PBS at 37 °C. Each syringe was filled with approximately 2 g of IBS, which was then extruded from the syringe manually at a constant speed [24,39].

2.3.4. SEM Analysis

The morphology and the internal porous structure of the lyophilized IBS samples were observed by using Scanning Electron Microscopy (SEM) (Zeiss, Evo LS10, Oberkochen, Germany) with 10 kV accelerating voltage at Yildiz Technical University, Istanbul, Turkey [30]. Samples were coated with gold-palladium before the experiment.

2.3.5. Rheological Measurements

Rheological measurements of IBS samples were performed by using a stress-controlled rheometer with a parallel plate geometry (diameter: 15 mm) (Anton Paar, MCR302, Graz, Austria) at Bogaziçi University, Istanbul, Turkey. The mechanical properties of the samples were measured at 0.1% strain and 10 rad/s frequency within the linear viscoelastic region of IBS. The oscillation amplitude and frequency sweep were carried out at 37 °C. A temperature sweep was performed from 15 to 45 °C at a heating rate of 2 °C/min in order to determine the gelation and setting temperatures. A time sweep was carried out at 37 °C to investigate the gelation kinetics. Finally, the shear thinning properties of IBS samples were analyzed both at 25 and 37 °C [24,40].

2.3.6. Compressive Strength Measurements

IBS samples were molded into columns of 7.0 mm of diameter and 10.0 mm length. The compressive strength of the samples was measured using a Geratech SH-500 (Geratech, Taiwan) and SH-20 testing device (Geratech, Taiwan) at Bogaziçi University, Istanbul, Turkey. The compressive strength of IBS samples was taken at 15% strain. Measurements were taken on days 1,3, 5 and 7 and 14 in an atmosphere of 100% humidity at 37 °C (n = 5) [41].

2.3.7. pH Changes

IBS samples were molded into discs, and after setting of the cement phase, they were immersed in PBS solution at 37 °C. The pH values of the PBS solution were measured at different time intervals including 5,10, 20, 30, 60, and 120 min, 1,3, 5, 7, 14, 21 and 28 days (n = 3) [41].

2.3.8. In Vitro Degradation

The degradation behavior of the IBS samples was measured in PBS at 37 °C. IBS samples were immersed in 12 well plates. At pre-determined times, the IBS samples were lyophilized and weighed. The remnant dry weight was calculated using Equation (1) where W_0 is the initial weight of the dry IBS samples and W_t is the dry weight of the IBS samples after t days of incubation [24].

$$\% = \frac{W_t}{W_0} \times 100 \; [\%] \tag{1}$$

3. Results and Discussion

3.1. Analysis of the Synthesized Powder

The TTCP/HA-based powder was synthesized as described in Section 2.2.2. Figure 1 shows the XRD analysis of the synthesized powder. The relevant Miller Indices of TTCP, HA and monetite are also presented.

Figure 1 shows that TTCP and hydroxyapatite peaks were detected since the powder was furnace-cooled [9,37,42,43]. A trace amount of monetite and calcium oxide (CaO) were also found from the XRD analysis. The matched peaks with JCPDS file No. 25-1137, No. 09-0432 and No. 09-0080 is given in Figure S1.

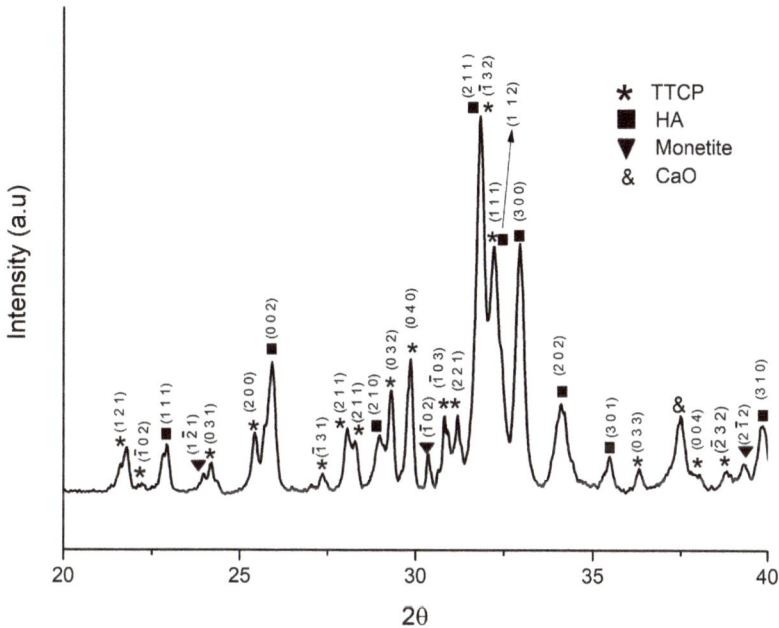

Figure 1. XRD (X-ray Diffraction) pattern of synthesized (tetracalcium phosphate/hydroxyapatite) TTCP/HA-based powder.

Figure 2 shows the FTIR spectrum of the synthesized powder.

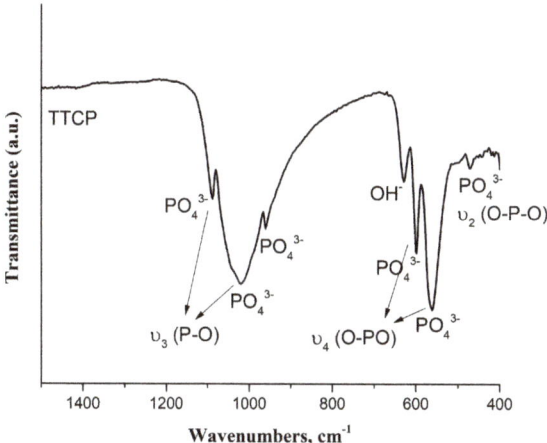

Figure 2. Fourier transform infrared (FTIR) spectrum of TTCP/HA-based powder by using KBr pellet method.

In FTIR spectrum, absorption peaks were found at 452, 471, 502, 567, 600, 628, 958, 987, 1044, 1091, 1625, 1922, 2001, 2077, 3435, 3570 and 3643 cm^{-1}. As labelled in Figure 2, PO_4^{3-} bands of TTCP were detected in the spectrum which were found similar with that of literature [43–45]. OH$^-$

stretching bands located at 628 and 3570 cm^{-1} indicated the presence of HA [42,46]. Absorption bands between 3000–3600 cm^{-1} indicated H$_2$O adsorbed. The peaks located at 1625, 1922, 2000 and 2077 cm^{-1} indicated the presence of carbonate content. These peaks were possibly found due to carbon dioxide absorption from the atmosphere [47]. Therefore, FTIR results supported the results observed from XRD spectrum which indicated synthesis of TTCP/HA-based powder [37,44].

3.2. Injectability of IBS Samples

Figure 3 shows that the mixture of bioceramic and polymeric components have a chewing gum-like consistency after mixing.

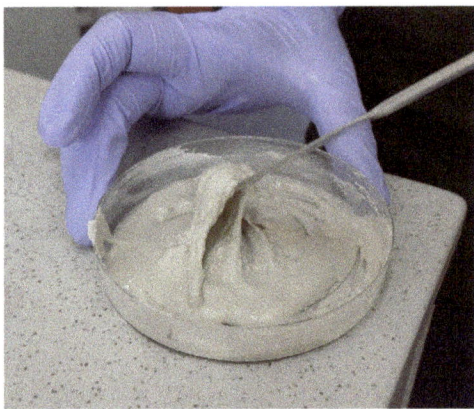

Figure 3. Chewing gum-like consistency of P50 samples after mixing of bioceramic and polymeric components.

Since the IBS samples had a chewing gum-like consistency, IBS samples can be molded into the desired shape of the complex bone defects. This consistency was achieved as a result of the presence of the liquid phase [12]. Figure S2 shows that all of the IBS samples possess cohesive stability and moldability. According to the extrusion videos, P30 and P50 samples had a higher stability than P20, as P20 pastes had a tendency to disintegrate during the extrusion process. Figure S3 shows that all IBS samples had high degree of injectability. Higher wt % of bioceramic components were also introduced into the polymeric component; however, these samples could not be extruded through 18-gauge. Therefore, the maximum wt % of bioceramic component was set at 50%.

3.3. Morphology of the IBS Samples

Figure 4 shows the morphologies of lyophilized hydrogels studied by SEM.

Figure 4 shows that all of the IBS samples had highly porous microstructure. When the wt % of the bioceramic component increased, the pore size decreased. P0 revealed that scaffolds had an interconnected, porous structure. SEM shows that bioceramics were well-adhered on the polymeric phase. Moreover, the bioceramic component was found to be homogenously distributed in the polymeric component [42,48] .

3.4. Rheological Measurements

Rheological measurements of IBS samples containing different wt % of the bioceramic component were evaluated. Figure 5 shows the amplitude sweep measurement of IBS samples.

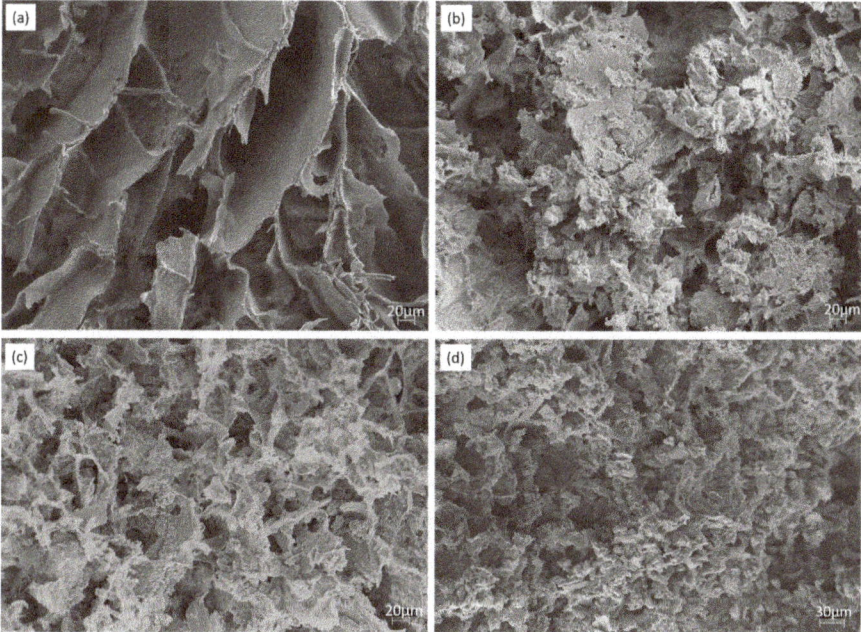

Figure 4. SEM images of (**a**) P0, (**b**) P20, (**c**) P30, (**d**) P50 IBS samples at a magnification of 500×.

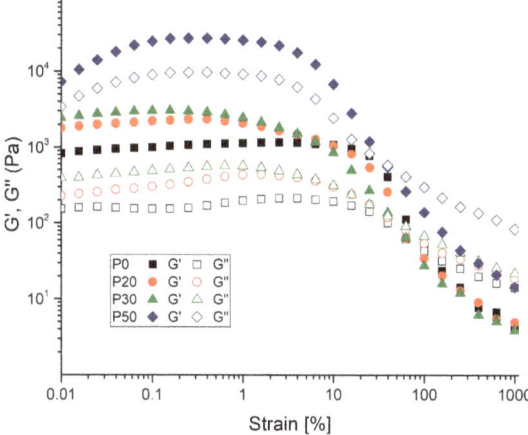

Figure 5. Amplitude-dependent variation of G' and G" changes of P0, P20, P30 and P50 IBS samples at a frequency of 10 rad/s at 37 °C (results expressed as mean ± standard error, n = 3).

Compared with P0 samples, other IBS samples present a broader linear viscoelastic region at 37 °C; as the strain required to break the network structure of IBS samples slightly increased. With the increase of wt % of the bioceramic component, the % strain required to break the network structure of IBS decreased. Therefore, in order to maintain the structural integrity of IBS samples, 0.1% strain was applied for frequency, temperature, and time sweep measurements. Figure 6 shows the frequency-dependent rheological results performed in the linear viscoelastic region under 0.1% strain.

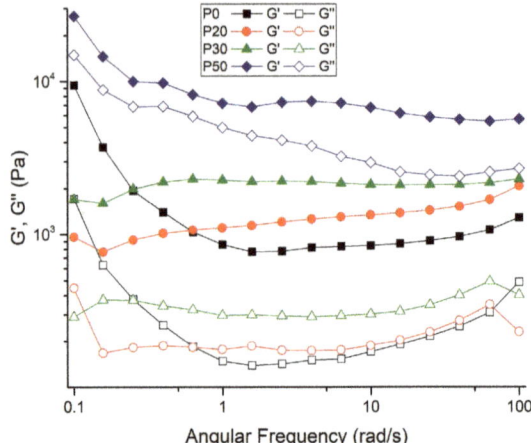

Figure 6. Frequency-dependent variation of G′ and G″ changes of P0, P20, P30 and P50 IBS samples at a 0.1% strain at 37 °C (results expressed as mean ± standard error, n = 3).

In the measured frequency range, all of the IBS samples had a higher storage modulus than the loss modulus, confirming the gelation and stabilization of their structure after setting at 37 °C [24,40]. The setting takes place in two stages. In the first stage, hardening occurs either by the hydration of the salts in the powder component or by a chelate reaction between MC and citric acid. At this time, the polymeric component and bioceramic components also have hydrogen bonds and ionic interactions. In the second stage of cement setting, the hardening occurs via the transformation of the bioceramic component to hydroxyapatite [9].

The rheological properties of the IBS samples were evaluated by the oscillatory rheometer as a function of temperature and time. Temperature and time sweep measurements were taken to examine the impact of bioceramic powder phase on the gelation and hardening mechanism. Figure 7 shows the temperature-dependent changes of G′ and G″ of IBS samples.

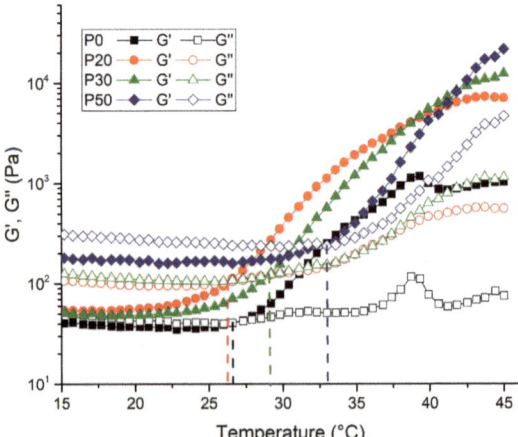

Figure 7. Temperature-dependent variation of G′ and G″ of IBS samples at 0.1% strain and 10 rad/s angular frequency (results expressed as mean ± standard error, n = 3).

The exponential increase of storage modulus with temperature implies the phase transition of the samples. The sol-gel transition temperature of the IBS samples shifted to a higher value with the addition of 30 and 50% of bioceramic mixture. The shifting of gelation temperature might be due to the change of the intra-molecular and inter-molecular interactions of MC chains [7]. The strong hydrogen bond between MC chains and CPC causes a change in the temperature sensitivity of the MC chains [49]. Figure 8 shows the gelation time at 37 °C.

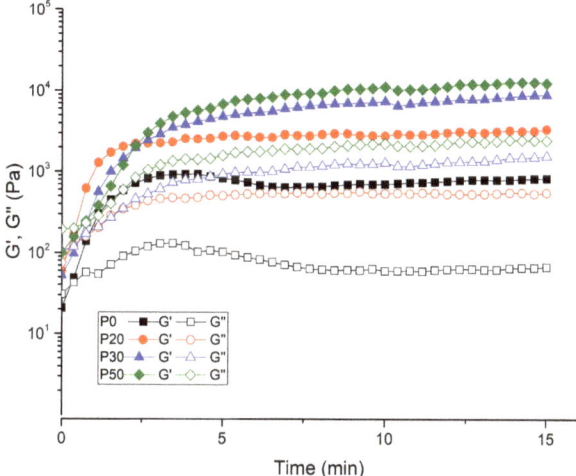

Figure 8. Time-dependent G' and G" of IBS samples measured at 37 °C, 0.1% strain, and 10 rad/s frequency (results expressed as mean ± standard error, n = 3).

Similar to the temperature sweep test, the exponential increase of the storage modulus with time at 37 °C suggests the hardening of IBS samples. The plateau point indicates the curing of the polymeric chains. The duration of curing decreases effectively with the increase of wt % of the bioceramic component [50–52].

Figure 9 shows the shear-rate dependent variation of viscosity for IBS samples both at 25 °C and 37 °C.

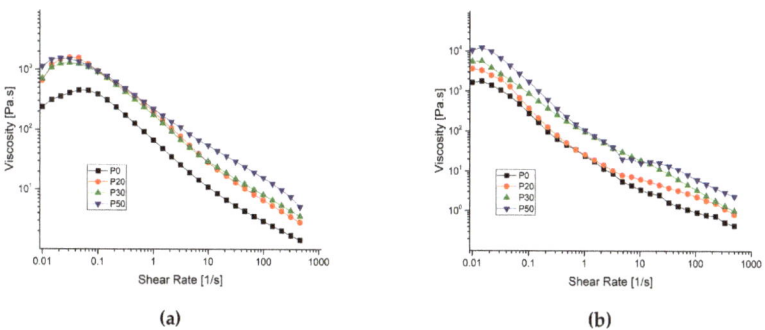

Figure 9. Shear-rate dependent variation of viscosity for IBS samples measured at (**a**) 25 °C and (**b**) 37 °C, respectively (results expressed as mean ± standard error, n = 3).

IBS samples were tested for their viscosity variation against change in the shear rate at 25 °C and 37 °C to observe whether IBS samples keep their injectability with respect to the increase of wt % of the bioceramic component. The results reveal that both at 25 °C and 37 °C, all of the samples had shear thinning properties and P0 had a considerably lower viscosity when compared to P20, P30 and P50 [24,41]. When the wt % of bioceramic component increased, viscosity also increased. This is due to the increase of resistance to flow as the number of the cement particles per unit volume increases [49].

3.5. pH Change

Figure 10 shows the pH change of the PBS after incubation of IBS samples at 37 °C.

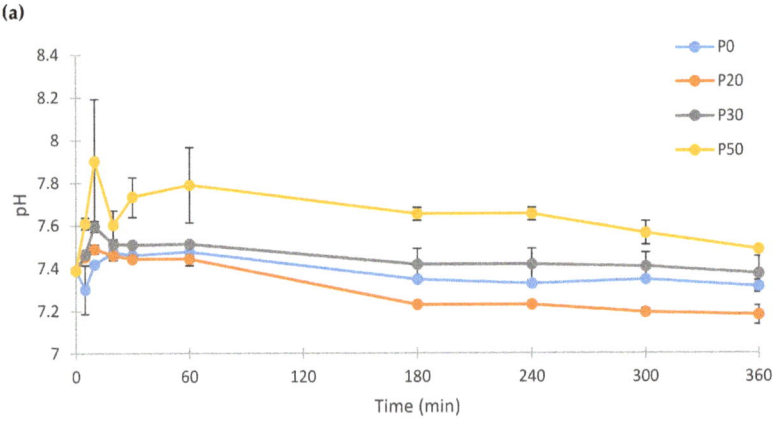

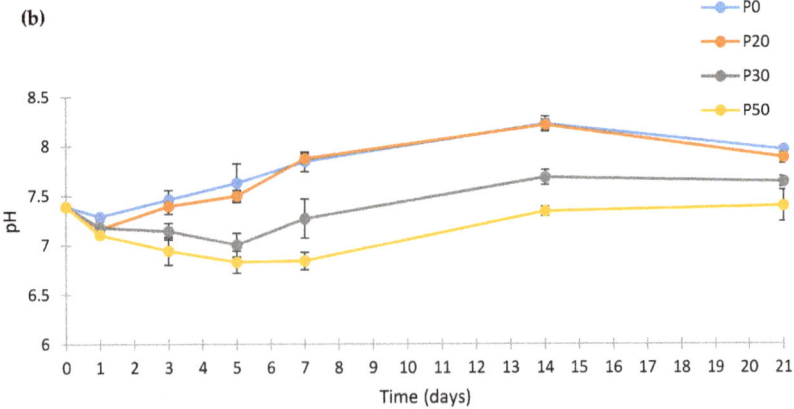

Figure 10. pH change of the PBS (phosphate buffer saline) that IBS samples incubated at 37 °C. (a) pH changes of IBS samples until 360 min and (b) pH changes of IBS samples until 21st day (results expressed as mean ± standard error, n = 3).

The pH profile of biomaterials in PBS is an important indicator of some of their possible biological responses. The pH response of the samples were measured for 21 days to monitor the pH changes after setting of IBS samples during the dissolution and re-precipitation. IBS samples had pH values between 7.89 and 7.39 at the end of day 21. Hence, it can be concluded that the prepared IBS samples may not cause any inflammatory reaction under biological conditions due to acidity. The pH of P50 samples was found to be higher than the other IBS samples at the beginning of incubation. P50 samples had the highest pH value until 60 min after which pH was gradually decreased. For P0, P20 and P30 IBS samples, the pH value reached a plateau after 3 h until the end of the first day. After day 1, pH values of P0, P20, and P30 started to increase slightly. The increase of pH after the 1st day of incubation was possibly due to the dissolution and transformation of TTCP into HA as indicated by Yokoyama et al. [41]. For P30 and P50 samples, the reduction of pH value until day 5 results in PO_4^{3-} consumption which leads to formation of an apatite-like phase [49].

3.6. In Vitro Degradation

Figure 11 shows % remaining weights of all IBS samples.

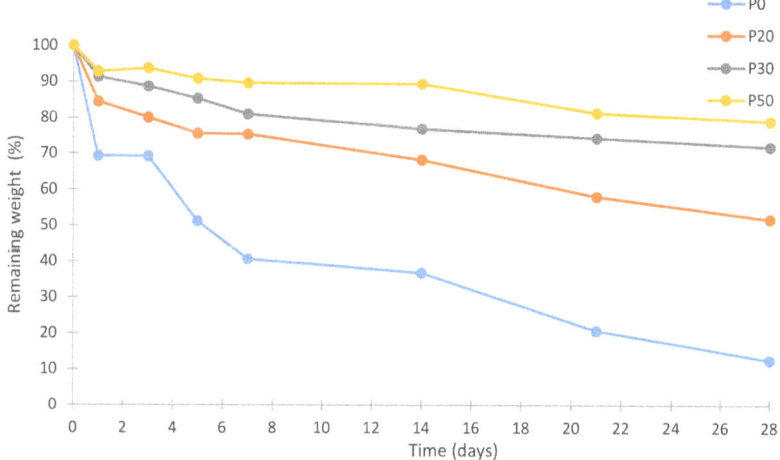

Figure 11. Remaining weights of IBS samples in PBS at 37 °C (results expressed as mean ± standard error, n = 3).

The in vitro degradation behavior of IBS samples was investigated by the measurement of % weight loss in PBS at 37 °C after the setting of the cement phase. After one week, P0 samples lost 60% of their weight due to the erosion of MC as Gupta et al. [53] and Tate et al. [54] reported. When the wt % of bioceramic powder component increased, the weight loss decreased. In vitro degradation studies were conducted without utilizing any enzymes. Therefore, a faster degradation rate of the IBS system is expected under in vivo conditions [55–57]. Therefore, the degradation rate of the IBS system may be further decreased with use of additives.

3.7. XRD Analysis

After the cement phase of the IBS samples was allowed to set, XRD analysis was conducted. Figure 12 shows the XRD patterns of the IBS samples.

As a result of these analyses, the peaks of powder components, TTCP, HA, DCPD, and CSD were observed from the XRD spectrum of P20, P30, and P50. The peak intensity of each IBS sample was increased as the wt % of the bioceramic powder increased. As Thai and Lee [9] concluded, initial XRD data did not reveal the setting mechanism of P20, P30, and P50 samples. Therefore, the XRD analysis was also performed after PBS studies to interpret the setting mechanism and mechanical behavior after incubation for 14 days. Figure 13 shows the XRD analysis after the incubation of IBS samples in PBS at 37 °C.

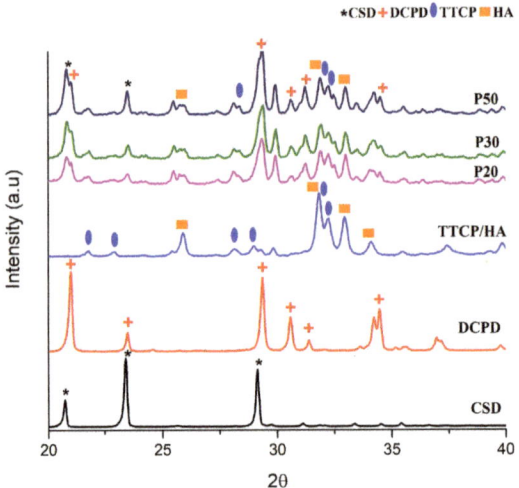

Figure 12. XRD analysis results of P20, P30 and P50 samples after setting.

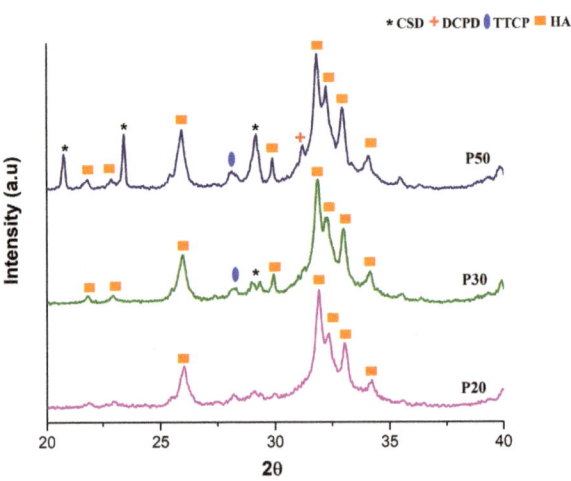

Figure 13. XRD analysis results of P20, P30 and P50 samples incubated in PBS at 37 °C for 14 days.

The deposition of an apatite layer on bone substitutes in the biological environment is an essential phenomenon as it indicates the osseointegration ability of implants [56]. The XRD results of the incubated IBS samples showed HA peaks which indicated the formation of an apatite layer on the samples. CSD peaks were also observed for P50 which is possibly due to a higher wt % of CSD present in P50 samples.

3.8. Compressive Strength Measurements

Figure 14 shows the compressive strength results of IBS samples in 100% humidity at 37 °C.

Figure 14 shows the compressive strength values of the samples after incubation in 100% humidity at 37 °C for 14 days. According to the results, P0 and P20 had almost the same compressive strength values.

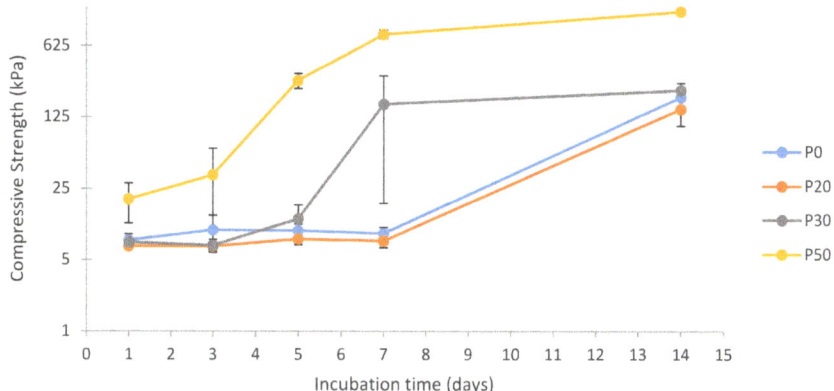

Figure 14. Log compressive strength of IBS samples vs. incubation time in 100% humidity at 37 °C (results expressed as mean ± standard error, n = 6).

Until day 7, the compressive strength of P30 samples had a similar trend with P0 and P20. However, interestingly, P30 had a significant increase in compressive strength on day 7. This increase was correlated with XRD results which indicated the phase transformation of TTCP into HA [41]. P50 samples had a much earlier rise in compressive strength than P30 samples; however, both P30 and P50 reached a plateau on day 7 indicating the completion of their phase transformation into HA. On day 14, P50 samples had approximately 7 times higher compressive strength when compared to P0, P20, and P30 samples. The compressive strength of cancellous bone varies between 0.22 to 10.44 MPa [4,58,59]. Compared to human cancellous bone, the compressive strength of IBS was found to be lower. One way to improve the mechanical properties of IBS is to increase the wt % of the bioceramic component. In this study, wt % of the bioceramic component could not be increased further due to the inability to inject IBS with higher wt % of the bioceramic component. The mechanical properties of IBS may be improved with the addition of carbon-based nanomaterials, such as carbon nanotubes and graphene oxide [60–62].

4. Conclusions

In this study, novel IBS were prepared by incorporation of different wt % of CaP/CS-based bioceramic powder into an MC-based solution. The rheological studies revealed that all of the samples had shear thinning properties; therefore, they had a high degree of injectability. This study showed that the incorporation of the bioceramic powder into MC-based polymeric matrices may improve the rheological, mechanical and degradation properties of IBS. In the future, it would be worthwhile to analyze the biocompatibility and biological responses of the developed IBS. Overall, the prepared IBS samples are promising candidates for the treatment of bone defects for non-load bearing applications.

Supplementary Materials: The following are available online at http://www.mdpi.com/1996-1944/11/4/604/s1, Figure S1: TTCP phase identified by XRD with JCPDS No. 25-1137, No. 09-0432 and No. 09-0080 files, Figure S2: Extrusion videos of IBS samples through syringe with 18-gauge needle. in PBS at 37 °C (**a**) extrusion of P0, (**b**) extrusion of P20, (**c**) extrusion of P30 and (**d**) extrusion of P50., Figure S3: Injectability of P0, P20, P30 and P50 samples.

Acknowledgments: The support of Boğaziçi University Research fund (Project No.: 12240) and TUBITAK (Project No.: 117M231) are kindly acknowledged.

Author Contributions: Öznur Demir Oğuz and Duygu Ege conceived and designed the experiments; Öznur Demir Oğuz performed the experiments; Öznur Demir Oğuz and Duygu Ege analyzed the data and wrote the paper.

Conflicts of Interest: The authors declare that they have no conflict of interest.

References

1. Brown, E.; Chow, L.C. A New Calcium Phosphate, Setting Cement. *J. Dent.* **1983**, *62*, 672.
2. Low, K.L.; Tan, S.H.; Zein, S.H.S.; Roether, J.A.; Mouriño, V.; Boccaccini, A.R. Calcium phosphate-based composites as injectable bone substitute materials. *J. Biomed. Mater. Res. Part B Appl. Biomater.* **2010**, *94*, 273–286. [CrossRef] [PubMed]
3. Ginebra, M.P.; Espanol, M.; Montufar, E.B.; Perez, R.A.; Mestres, G. New processing approaches in calcium phosphate cements and their applications in regenerative medicine. *Acta Biomater.* **2010**, *6*, 2863–2873. [CrossRef] [PubMed]
4. O'Neill, R.; McCarthy, H.O.; Montufar, E.B.; Ginebra, M.P.; Wilson, D.I.; Lennon, A.; Dunne, N. Critical review: Injectability of calcium phosphate pastes and cements. *Acta Biomater.* **2017**, *50*, 1–19. [CrossRef] [PubMed]
5. Lewis, G. Injectable bone cements for use in vertebroplasty and kyphoplasty: state-of-the-art review. *J. Biomed. Mater. Res. B Appl. Biomater.* **2006**, *76*, 456–468. [CrossRef] [PubMed]
6. Habraken, W.; Habibovic, P.; Epple, M.; Bohner, M. Calcium phosphates in biomedical applications: Materials for the future? *Mater. Today* **2016**, *19*, 69–87. [CrossRef]
7. Chow, L.C. Next generation calcium phosphate-based biomaterials. *Dent. Mater. J.* **2009**, *28*, 1–10. [CrossRef] [PubMed]
8. Barrère, F.; van Blitterswijk, C.A.; de Groot, K. Bone regeneration: Molecular and cellular interactions with calcium phosphate ceramics. *Int. J. Nanomed.* **2006**, *1*, 317–332.
9. Thai, V.V.; Lee, B.T. Fabrication of calcium phosphate-calcium sulfate injectable bone substitute using hydroxy-propyl-methyl-cellulose and citric acid. *J. Mater. Sci. Mater. Med.* **2010**, *21*, 1867–1874. [CrossRef] [PubMed]
10. Rangabhatla, A.S.L.; Tantishaiyakul, V.; Oungbho, K.; Boonrat, O. Fabrication of pluronic and methylcellulose for etidronate delivery and their application for osteogenesis. *Int. J. Pharm.* **2016**, *499*, 110–118. [CrossRef] [PubMed]
11. Chen, Z.; Zhang, X.; Kang, L.; Xu, F.; Wang, Z.; Cui, F.-Z.; Guo, Z. Recent progress in injectable bone repair materials research. *Front. Mater. Sci.* **2015**, *9*, 332–345. [CrossRef]
12. Perez, R.A.; Shin, S.-H.; Han, C.-M.; Kim, H.-W. Bioactive injectables based on calcium phosphates for hard tissues: A recent update. *Tissue Eng. Regen. Med.* **2015**, *12*, 143–153. [CrossRef]
13. Kondiah, P.J.; Choonara, Y.E.; Kondiah, P.P.D.; Marimuthu, T.; Kumar, P.; Du Toit, L.C.; Pillay, V. A review of injectable polymeric hydrogel systems for application in bone tissue engineering. *Molecules* **2016**, *21*. [CrossRef] [PubMed]
14. Wang, L.; Zhang, C.; Li, C.; Weir, M.D.; Wang, P.; Reynolds, M.A.; Zhao, L.; Xu, H.H.K. Injectable calcium phosphate with hydrogel fibers encapsulating induced pluripotent, dental pulp and bone marrow stem cells for bone repair. *Mater. Sci. Eng. C* **2016**, *69*, 1125–1136. [CrossRef] [PubMed]
15. Priya, M.V.; Sivshanmugam, A.; Boccaccini, A.R.; Goudouri, O.M.; Sun, W.; Hwang, N.; Deepthi, S.; Nair, S.V.; Jayakumar, R. Injectable osteogenic and angiogenic nanocomposite hydrogels for irregular bone defects Injectable osteogenic and angiogenic nanocomposite hydrogels for irregular bone defects. *Biomed. Mater.* **2016**, *11*. [CrossRef]
16. Shimokawa, K.; Saegusa, K.; Ishii, F. Rheological properties of reversible thermo-setting in situ gelling solutions with the methylcellulose-polyethylene glycol-citric acid ternary system (2): Effects of various water-soluble polymers and salts on the gelling temperature. *Colloid Surf. B Biointerfaces* **2009**, *74*, 56–58. [CrossRef] [PubMed]
17. Basnett, P.; Knowles, J.C.; Pishbin, F.; Smith, C.; Keshavarz, T.; Boccaccini, A.R.; Roy, I. Novel biodegradable and biocompatible poly(3-hydroxyoctanoate)/bacterial cellulose composites. *Adv. Eng. Mater.* **2012**, *14*, 330–343. [CrossRef]
18. Perale, G.; Rossi, F.; Santoro, M.; Peviani, M.; Papa, S.; Llupi, D.; Torriani, P.; Micotti, E.; Previdi, S.; Cervo, L.; et al. Multiple drug delivery hydrogel system for spinal cord injury repair strategies. *J. Control. Release* **2012**, *159*, 271–280. [CrossRef] [PubMed]

19. Bain, M.K.; Maity, D.; Bhowmick, B.; Mondal, D.; Mollick, M.M.R.; Sarkar, G.; Bhowmik, M.; Rana, D.; Chattopadhyay, D. Effect of PEG-salt mixture on the gelation temperature and morphology of MC gel for sustained delivery of drug. *Carbohydr. Polym.* **2013**, *91*, 529–536. [CrossRef] [PubMed]
20. Jeong, N.; Park, J.; Yoo, K.; Kim, W.; Kim, D.H.; Yoon, S.Y. Preparation, characterization, and in-vitro performance of novel injectable silanized-hydroxypropyl methylcellulose/phase-transformed calcium phosphate composite bone cements. *Curr. Appl. Phys.* **2016**, *16*, 1523–1532. [CrossRef]
21. Ghanaati, S.; Barbeck, M.; Hilbig, U.; Hoffmann, C.; Unger, R.E.; Sader, R.A.; Peters, F.; Kirkpatrick, C.J. An injectable bone substitute composed of beta-tricalcium phosphate granules, methylcellulose and hyaluronic acid inhibits connective tissue influx into its implantation bed in vivo. *Acta Biomater.* **2011**, *7*, 4018–4028. [CrossRef] [PubMed]
22. Krause, M.; Oheim, R.; Catala-Lehnen, P.; Pestka, J.M.; Hoffmann, C.; Huebner, W.; Peters, F.; Barvencik, F.; Amling, M. Metaphyseal bone formation induced by a new injectable beta-TCP-based substitute: A controlled study in rabbits. *J. Biomater. Appl.* **2014**, *28*, 859–868. [CrossRef] [PubMed]
23. Patenaude, M.; Hoare, T. Injectable, mixed natural-synthetic polymer hydrogels with modular properties. *Biomacromolecules* **2012**, *13*, 369–378. [CrossRef] [PubMed]
24. Liu, Z.; Yao, P. Injectable thermo-responsive hydrogel composed of xanthan gum and methylcellulose double networks with shear-thinning property. *Carbohydr. Polym.* **2015**, *132*, 490–498. [CrossRef] [PubMed]
25. Félix Lanao, R.P.; Sariibrahimoglu, K.; Wang, H.; Wolke, J.G.C.; Jansen, J.A.; Leeuwenburgh, S.C.G. Accelerated calcium phosphate cement degradation due to incorporation of glucono-delta-lactone microparticles. *Tissue Eng. Part A* **2014**, *20*, 378–388. [CrossRef] [PubMed]
26. Dessì, M.; Alvarez-Perez, M.A.; De Santis, R.; Ginebra, M.P.; Planell, J.A.; Ambrosio, L. Bioactivation of calcium deficient hydroxyapatite with foamed gelatin gel. A new injectable self-setting bone analogue. *J. Mater. Sci. Mater. Med.* **2014**, *25*, 283–295. [CrossRef] [PubMed]
27. Utech, S.; Boccaccini, A.R. A review of hydrogel-based composites for biomedical applications: enhancement of hydrogel properties by addition of rigid inorganic fillers. *J. Mater. Sci.* **2016**, *51*, 271–310. [CrossRef]
28. Bongio, M.; Nejadnik, M.R.; Kasper, F.K.; Mikos, A.G.; Jansen, J.A.; Leeuwenburgh, S.C.G.; van den Beucken, J.J.J.P. Development of an in vitro confinement test to predict the clinical handling of polymer-based injectable bone substitutes. *Polym. Test.* **2013**, *32*, 1379–1384. [CrossRef]
29. Nishinari, K.; Hofmann, K.E.; Kohyama, K.; Moritaka, H.; Nishinari, N.; Watase, M. Polysaccharide-protein interaction: A rheological study of the gel-sol transition of a gelatin-methylcellulose-water system. *Biorheology* **1993**, *30*, 243–252. [CrossRef] [PubMed]
30. Demitri, C.; Del Sole, R.; Scalera, F.; Sannino, A.; Vasapollo, G.; Maffezzoli, A.; Ambrosio, L.; Nicolais, L. Novel superabsorbent cellulose-based hydrogels crosslinked with citric acid. *J. Appl. Polym. Sci.* **2008**, *110*, 2453–2460. [CrossRef]
31. Habraken, W.J.E.M.; Jonge, L.T.; De Wolke, J.G.C.; Yubao, L.; Mikos, A.G.; Jansen, J.A. Introduction of gelatin microspheres into an injectable calcium phosphate cement. *J. Biomed. Mater. Res. A* **2008**, *87*, 643–655. [CrossRef] [PubMed]
32. Sanmartín-Masiá, E.; Poveda-Reyes, S.; Gallego Ferrer, G. Extracellular matrix–inspired gelatin/hyaluronic acid injectable hydrogels. *Int. J. Polym. Mater. Polym. Biomater.* **2017**, *66*, 280–288. [CrossRef]
33. Tang, Y.; Wang, X.; Li, Y.; Lei, M.; Du, Y.; Kennedy, J.F.; Knill, C.J. Production and characterisation of novel injectable chitosan/methylcellulose/salt blend hydrogels with potential application as tissue engineering scaffolds. *Carbohydr. Polym.* **2010**, *82*, 833–841. [CrossRef]
34. Sadiasa, A.; Sarkar, S.K.; Franco, R.A.; Min, Y.K.; Lee, B.T. Bioactive glass incorporation in calcium phosphate cement-based injectable bone substitute for improved in vitro biocompatibility and in vivo bone regeneration. *J. Biomater. Appl.* **2013**, *28*, 739–756. [CrossRef] [PubMed]
35. Hempel, U.; Reinstorf, A.; Poppe, M.; Fischer, U.; Gelinsky, M.; Pompe, W.; Wenzel, K.W. Proliferation and Differentiation of Osteoblasts on Biocement D Modified with Collagen Type I and Citric Acid. *J. Biomed. Mater. Res. Part B Appl. Biomater.* **2004**, *71B*, 130–143. [CrossRef] [PubMed]
36. Wang, X.; Ye, J.; Wang, H. Effects of Additives on the Rheological Properties and Injectability of a Calcium Phosphate Bone Substitute Material. *J. Biomed. Mater. Res. Part B Appl. Polym.* **2005**, *78B*, 259–264. [CrossRef] [PubMed]
37. Guo, D.; Xu, K.; Han, Y. Influence of cooling modes on purity of solid-state synthesized tetracalcium phosphate. *Mater. Sci. Eng. B Solid-State Mater. Adv. Technol.* **2005**, *116*, 175–181. [CrossRef]

38. Song, H.Y.; Rahman, A.H.M.E.; Lee, B.T. Fabrication of calcium phosphate-calcium sulfate injectable bone substitute using chitosan and citric acid. *J. Mater. Sci. Mater. Med.* **2009**, *20*, 935–941. [CrossRef] [PubMed]
39. Alves, H.L.R.; dos Santos, L.A.; Bergmann, C.P. Injectability evaluation of tricalcium phosphate bone cement. *J. Mater. Sci. Mater. Med.* **2008**, *19*, 2241–2246. [CrossRef] [PubMed]
40. Fan, R.R.; Deng, X.H.; Zhou, L.X.; Gao, X.; Fan, M.; Wang, Y.L.; Guo, G. Injectable thermosensitive hydrogel composite with surface-functionalized calcium phosphate as raw materials. *Int. J. Nanomed.* **2014**, *9*, 615–626. [CrossRef]
41. Yokoyama, A.; Yamamoto, S.; Kawasaki, T.; Kohgo, T.; Nakasu, M. Development of calcium phosphate cement using chitosan and citric acid for bone substitute materials. *Biomaterials* **2002**, *23*, 1091–1101. [CrossRef]
42. Huang, Z.; Feng, Q.; Yu, B.; Li, S. Biomimetic properties of an injectable chitosan / nano-hydroxyapatite / collagen composite. *Mater. Sci. Eng. C* **2011**, *31*, 683–687. [CrossRef]
43. Radwan, M.M.; Abd El-Hamid, H.K.; Nagi, S.M. Synthesis, properties and hydration characteristics of novel nano-size mineral trioxide and tetracalcium phosphate for dental applications. *Orient. J. Chem.* **2016**, *32*, 2459–2472. [CrossRef]
44. Jayasree, R.; Kumar, T.S.; Kavya, K.P.S.; Nankar, P.R.; Mukesh, D. Self Setting Bone Cement Formulations Based on Egg shell Derived TetraCalcium Phosphate BioCeramics. *Bioceram. Dev. Appl.* **2015**, *5*, 1–6. [CrossRef]
45. Liao, J.; Duan, X.; Li, Y.; Zheng, C.; Yang, Z.; Zhou, A.; Zou, D. Synthesis and mechanism of tetracalcium phosphate from nanocrystalline precursor. *J. Nanomater.* **2014**, *2014*. [CrossRef]
46. Kim, H.; Camata, R.P.; Vohra, Y.K.; Lacefield, W.R. Control of phase composition in hydroxyapatite/tetracalcium phosphate biphasic thin coatings for biomedical applications. *J. Mater. Sci. Mater. Med.* **2005**, *16*, 961–966. [CrossRef] [PubMed]
47. Eslami, H.; Solati-Hashjin, M.; Tahriri, M. Synthesis and Characterization of Hydroxyapatite Nanocrystals via Chemical Precipitation Technique. *Iran. J. Pharm. Sci.* **2008**, *4*, 127–134.
48. Liu, W.; Zhang, J.; Rethore, G.; Khairoun, K.; Pilet, P.; Tancret, F.; Bouler, J.M.; Weiss, P. A novel injectable, cohesive and toughened Si-HPMC (silanized-hydroxypropyl methylcellulose) composite calcium phosphate cement for bone substitution. *Acta Biomater.* **2014**, *10*, 3335–3345. [CrossRef] [PubMed]
49. Marefat Seyedlar, R.; Nodehi, A.; Atai, M.; Imani, M. Gelation behavior of in situ forming gels based on HPMC and biphasic calcium phosphate nanoparticles. *Carbohydr. Polym.* **2014**, *99*, 257–263. [CrossRef] [PubMed]
50. Ghorbani, M.; Ai, J.; Nourani, M.R.; Azami, M.; Hashemi Beni, B.; Asadpour, S.; Bordbar, S. Injectable natural polymer compound for tissue engineering of intervertebral disc: In vitro study. *Mater. Sci. Eng. C* **2017**, *80*, 502–508. [CrossRef] [PubMed]
51. Arvidson, S.A.; Lott, J.R.; McAllister, J.W.; Zhang, J.; Bates, F.S.; Lodge, T.P.; Sammler, R.L.; Li, Y.; Brackhagen, M. Interplay of phase separation and thermoreversible gelation in aqueous methylcellulose solutions. *Macromolecules* **2013**, *46*, 300–309. [CrossRef]
52. Wu, J.; Liu, J.; Shi, Y.; Wan, Y. Rheological, mechanical and degradable properties of injectable chitosan/silk fibroin/hydroxyapatite/glycerophosphate hydrogels. *J. Mech. Behav. Biomed. Mater.* **2016**, *64*, 161–172. [CrossRef] [PubMed]
53. Gupta, D.; Tator, C.H.; Shoichet, M.S. Fast-gelling injectable blend of hyaluronan and methylcellulose for intrathecal, localized delivery to the injured spinal cord. *Biomaterials* **2006**, *27*, 2370–2379. [CrossRef] [PubMed]
54. Tate, M.C.; Shear, D.A.; Hoffman, S.W.; Stein, D.G.; LaPlaca, M.C. Biocompatibility of methylcellulose-based constructs designed for intracerebral gelation following experimental traumatic brain injury. *Biomaterials* **2001**, *22*, 1113–1123. [CrossRef]
55. Ma, X.; Zhang, L.; Fan, D.; Xue, W.; Zhu, C.; Li, X.; Liu, Y.; Liu, W.; Ma, P.; Wang, Y. Physicochemical properties and biological behavior of injectable crosslinked hydrogels composed of pullulan and recombinant human-like collagen. *J. Mater. Sci.* **2017**, *52*, 3771–3785. [CrossRef]
56. Ding, Y.; Tang, S.; Yu, B.; Yan, Y.; Li, H.; Wei, J.; Su, J. In vitro degradability, bioactivity and primary cell responses to bone cements containing mesoporous magnesium–calcium silicate and calcium sulfate for bone regeneration. *J. R. Soc. Interface* **2015**, *12*, 20150779. [CrossRef] [PubMed]

57. Qasim, S.B.; Husain, S.; Huang, Y.; Pogorielov, M.; Deineka, V.; Lyndin, M.; Rawlinson, A.; Rehman, I.U. In-vitro and in-vivo degradation studies of freeze gelated porous chitosan composite scaffolds for tissue engineering applications. *Polym. Degrad. Stab.* **2017**, *136*, 31–38. [CrossRef]
58. Misch, C.E.; Qu, Z.; Bidez, M.W. Mechanical properties of trabecular bone in the human mandible: Implications for dental implant treatment planning and surgical placement. *J. Oral Maxillofac. Surg.* **1999**, *57*, 700–706. [CrossRef]
59. Baino, F. Ceramics for bone replacement. In *Advances in Ceramic Biomaterials*; Elsevier: Amsterdam, The Netherlands, 2017; pp. 249–278. ISBN 9780081008812.
60. Ege, D.; Kamali, A.R.; Boccaccini, A.R. Graphene Oxide/Polymer-Based Biomaterials. *Adv. Eng. Mater.* **2017**, *19*. [CrossRef]
61. Newman, P.; Minett, A.; Ellis-Behnke, R.; Zreiqat, H. Carbon nanotubes: Their potential and pitfalls for bone tissue regeneration and engineering. *Nanomed. Nanotechnol. Biol. Med.* **2013**, *9*, 1139–1158. [CrossRef] [PubMed]
62. Gholami, F.; Zein, S.H.S.; Gerhardt, L.-C.; Low, K.L.; Tan, S.H.; McPhail, D.S.; Grover, L.M.; Boccaccini, A.R. Cytocompatibility, bioactivity and mechanical strength of calcium phosphate cement reinforced with multi-walled carbon nanotubes and bovine serum albumin. *Ceram. Int.* **2013**, *39*, 4975–4983. [CrossRef]

© 2018 by the authors. Licensee MDPI, Basel, Switzerland. This article is an open access article distributed under the terms and conditions of the Creative Commons Attribution (CC BY) license (http://creativecommons.org/licenses/by/4.0/).

Article

The Mechanical Properties of Biocompatible Apatite Bone Cement Reinforced with Chemically Activated Carbon Fibers

Anne V. Boehm [1], Susanne Meininger [2], Annemarie Tesch [1], Uwe Gbureck [2] and Frank A. Müller [1],*

[1] Otto Schott Institute of Materials Research (OSIM), Friedrich Schiller University Jena, Löbdergraben 32, 07743 Jena, Germany; anne.boehm@uni-jena.de (A.V.B.); annemarie.tesch@uni-jena.de (A.T.)
[2] Department for Functional Materials in Medicine and Dentistry (FMZ), University of Würzburg, Pleicherwall 2, 97070 Würzburg, Germany; susanne.meininger@fmz.uni-wuerzburg.de (S.M.); uwe.gbureck@fmz.uni-wuerzburg.de (U.G.)
* Correspondence: frank.mueller@uni-jena.de; Tel.: +49-3641-947-750

Received: 14 December 2017; Accepted: 24 January 2018; Published: 26 January 2018

Abstract: Calcium phosphate cement (CPC) is a well-established bone replacement material in dentistry and orthopedics. CPC mimics the physicochemical properties of natural bone and therefore shows excellent in vivo behavior. However, due to their brittleness, the application of CPC implants is limited to non load bearing areas. Generally, the fiber-reinforcement of ceramic materials enhances fracture resistance, but simultaneously reduces the strength of the composite. Combining strong C-fiber reinforcement with a hydroxyapatite to form a CPC with a chemical modification of the fiber surface allowed us to adjust the fiber–matrix interface and consequently the fracture behavior. Thus, we could demonstrate enhanced mechanical properties of CPC in terms of bending strength and work of fracture to a strain of 5% (WOF5). Hereby, the strength increased by a factor of four from 9.2 ± 1.7 to 38.4 ± 1.7 MPa. Simultaneously, the WOF5 increased from 0.02 ± 0.004 to 2.0 ± 0.6 kJ·m^{-2}, when utilizing an *aqua regia*/$CaCl_2$ pretreatment. The cell proliferation and activity of MG63 osteoblast-like cells as biocompatibility markers were not affected by fiber addition nor by fiber treatment. CPC reinforced with chemically activated C-fibers is a promising bone replacement material for load-bearing applications.

Keywords: calcium phosphate cement; damage tolerant cement; carbon fiber reinforcement; interface control; fiber–matrix interaction

---

1. Introduction

The chemical and crystallographic similarity of hydroxyapatite (HAp) to bone apatite mineral found in mammalian hard tissues has made this material most attractive for replacing human bones and teeth [1–4]. Calcium phosphate cement (CPC) for bone regeneration applications has been under intense research for more than 30 years now [5]. Numerous formulations have been examined involving tailoring of the degradability, application time, and mechanical properties of CPC [6–8]. Commercial products are already in use for non-load bearing applications, like craniofacial or maxillodental surgery [7,9,10]. However, the main problem of CPC is related to its brittleness, resulting in a low fracture toughness [11]. To overcome this problem, reinforced CPC has been investigated using various reinforcement strategies [12,13], with particles or whiskers [14,15], short fibers for e.g., injectable CPC [16,17], and long fibers [18]. In a previous work on different fiber reinforcements, carbon fibers (C-fibers) proved their suitability for enhancing the strength and work of fracture (WOF) of CPC [8]. Xu et al. [18,19] reported on the mechanics of C-fiber reinforced CPC composites. They found that

5.7 vol % of C-fibers with a length of 75 mm in apatite cement resulted in a bending strength and WOF of 59 ± 11 MPa and 6.6 ± 1.2 kJ·m^{-2}, respectively, when compared to pure CPC with a strength of 13 ± 3 MPa and a WOF of 0.04 ± 0.01 kJ·m^{-2}. The main reinforcement mechanism was found to be the pull-out of fibers over long distances, which results from the weak interfacial strength between the matrix and continuous fiber reinforcement.

Although the strengthening and toughening of CPC have already been achieved through the use of C-fibers, to the best of our knowledge, the surface functionalization of the fibers has not yet been considered, and therefore the full potential of this reinforcement strategy is still not exploited. Adjusting the fiber–matrix interface by activating the fibers' surface alters the wetting and precipitation process and will consequently influence the mechanical properties and the failure mechanism. In the present work, we focus on the mechanical properties of CPC reinforced with C-fibers that were activated by utilizing different oxidation agents, followed by a calcium adhesion process. Since the addition of untreated fibers and particularly the chemical pretreatment of the fibers can have a decisive impact on the cellular behavior of these composites, their biocompatibility was tested with the osteoblast-like cell line MG63.

2. Results and Discussion

2.1. Fiber Activation

Modification of the fibers by hydrogen peroxide (H_2O_2) or *aqua regia* and the subsequent calcium adhesion (*aqua regia*/$CaCl_2$) was characterized by X-ray photoelectron spectroscopy (XPS) and contact angle measurements, as shown in Figure 1 and Table 1, respectively. Spectra of the C1s and O1s XPS peaks were measured. Due to oxygen adsorption during exposure to air as well as during fiber processing, starting from the precursor polymer polyacrylonitrile (PAN), oxygen is hence present in all fibers, regardless of modification. Beyond oxygen absorption, oxide lattice oxygen was also detected at binding energy (BE) 529–530 eV as a side effect of fiber processing [20].Therefore, the interpretation of the quantity of oxygen or the oxygen to carbon ratio is disputable in this case. However, by analyzing the signals with single-peak fitting (blue curves in C1s spectra, red in O1s spectra), different bond types can be assigned to the binding energies (BEs) and therefore the surface chemistry can be discussed.

The predominant species found in the C1s spectra (Figure 1a–c) of all fiber treatments are aromatic or aliphatic carbon, which are also known for graphite at binding energies (BEs) of 284.4 eV [21]. Additionally, untreated fibers and H_2O_2-oxidized fibers evolve signals at BE = 285.9–286.0 eV which can be attributed to hydroxyl groups (–OH) [22]. In the case of the *aqua regia*/$CaCl_2$ treatment, the spectra shift to a higher BE with a peak at 288.8 eV occurs, which can be attributed to carboxyl groups (–COOH) [23]. Furthermore, small quantities of adsorbed carbon oxide (CO) or carbon dioxide (CO_2) were detected for all fibers at a BE of roughly 291 eV [24].

Focusing on the O1s spectra (Figure 1d–f), only small differences between untreated fibers and H_2O_2 modification occurred. In those cases, the predominant oxygen species is found at a BE of 532.2–532.5 eV which can be attributed to –OH [22]. In contrast, oxidation using *aqua regia* resulted in a shift of the O1s signal towards lower BE, namely an additional peak at 531.0 eV which can be assigned to carbonyl groups (–CO) [21,22,24]. Furthermore, for all fiber types, but with different relative intensity, carboxyl groups (–COOH) were detected at BEs of 534.2–534.9 eV [21,22,24]. Hereby, the relative intensity, referring to the O1s intensity, is related to the fraction of this group. An increased ratio of –COOH was observed in the case of an *aqua regia*/$CaCl_2$ treatment.

Combining both spectra, several types of oxygen binding can be assigned. In the case of untreated and H_2O_2-pretreated fibers, mostly adsorbed oxygen and –OH groups are present, whereas the treatment with *aqua regia* causes the occurrence of –CO and –COOH. Considering that the setting reaction proceeds under basic conditions, a partial dissociation of those groups is possible. Taking into account pK_a-values of 9–11 for –OH [21,22], only a very minor fraction of –OH groups is deprotonated and hence no calcium adhesion was found in the XPS spectra for untreated and H_2O_2-treated fibers.

On the contrary, the lower pK_a values of ~2–4 for the hydration of carbonyl compounds with hydroxy ions (OH$^-$) [23,24] result in significant amounts of deprotonated –OH, which, like deprotonated carboxylic acid groups (COO$^-$), have a strong tendency to bind calcium. Therefore, XPS proved calcium adhesion to the *aqua regia*-pretreated C-fibers.

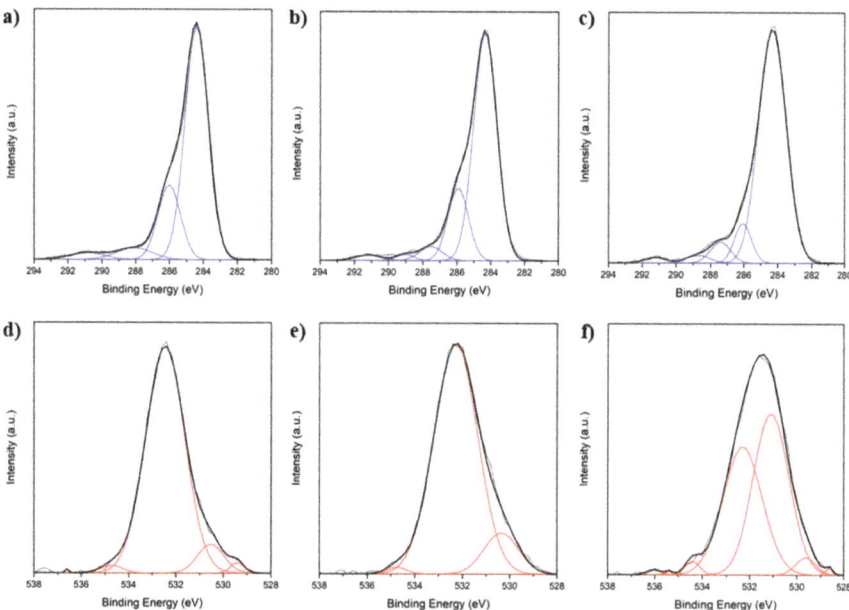

Figure 1. C1s (**a–c**) and O1s (**d–f**) measured by X-ray photoelectron spectroscopy for (**a,d**) untreated; (**b,e**) H$_2$O$_2$-treated; and (**c,f**) *aqua regia*/CaCl$_2$-treated C-fibers.

In Table 1 the water contact angles of untreated and chemically modified C-fibers are shown. The fiber oxidation leads to a significant decrease of the contact angles from 71° for untreated fibers to 64° for H$_2$O$_2$-treated and to 62° for *aqua regia*/CaCl$_2$-treated fibers. Reduced contact angles represent enhanced wettability, which was also observed during the handling and manufacturing of CPC composite samples.

2.2. Cement Setting

The setting reaction from α-tricalcium phosphate (α-TCP) to calcium deficient hydroxy apatite (CDHA) was characterized using X-ray diffraction (XRD) and SEM (Figure 2). Hereby, the plate-shaped precipitation of CDHA was confirmed. The kinetics of the setting reaction were investigated using the Gilmore needle test (Table 1). Both initial and final setting time are reduced significantly by inducing fibers. The reduced initial and final setting times are preferable for medical applications, due to time savings in the surgery process [25]. Especially in the case of the initial setting time, the reduction can be attributed to fibers acting as heterogenous seeds for the precipitation. The strongest reduction was observed in the case of *aqua regia*/CaCl$_2$ treatment, where fibers had already bound Ca^{2+} ions. Thus, the first step for apatite nucleation has already occurred and PO$_4^{3-}$ is directly attracted to such sites. Consequently, the crystal growth is accelerated the most. This effect is also represented by the final setting time, where a shortening by 50% was observed for the composite. Since the cement morphology around the fibers changed, one can assume that the precipitation process varies for different fiber modifications (Figure 3). Whereas untreated fibers show almost no interaction with the CPC matrix (Figure 3b), crystal growth can be observed on H_2O_2-treated (Figure 3c) as well as on *aqua regia*/$CaCl_2$-treated fibers (Figure 3d). In the latter case the crystals appear bigger than for the H_2O_2 treatment, which indicates a faster crystal growth or a slower nucleation rate. It is of value to mention that due to the acidic character of the fiber surface, not only the calcium phosphate nucleation process is affected or altered, but the chemical composition of the precipitate could also differ. Under the given conditions, besides CDHA, either brushite or octacalcium phosphate (OCP) could be precipitated. However, due to the addition of sodium citrate, the precipitation of brushite is inhibited, and therefore its formation seems unlikely [26,27]. On the other hand, the formation of fast growing OCP crystals is possible at high supersaturation on an acidic fiber surface [28]. In conclusion, two explanations for the reduced setting time in the case of an *aqua regia*/$CaCl_2$ treatment are possible. Either the setting time is reduced by an accelerated crystal growth due to the presence of calcium binding groups and a calcium pre-saturation on the fiber surface, or the change of precipitation product from CDHA to OCP, which shows faster crystal growth than CDHA may be a factor. However, the formation of OCP is only likely in close proximity to the fibers' surface, but not in the setting of the CPC matrix that consists of CDHA.

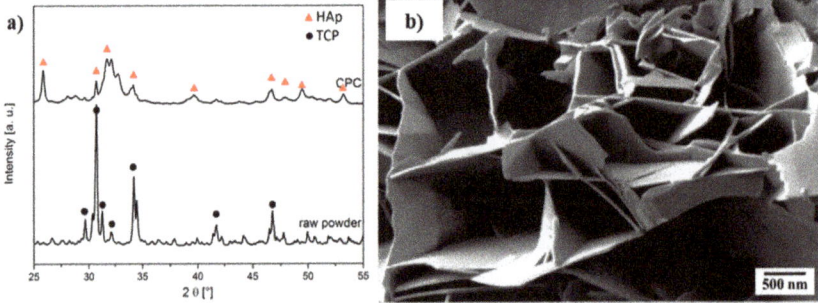

Figure 2. Characterization of (**a**) the phase composition of raw powders and set cement using X-ray diffraction and (**b**) the morphology of the cement matrix by scanning electron microscopy.

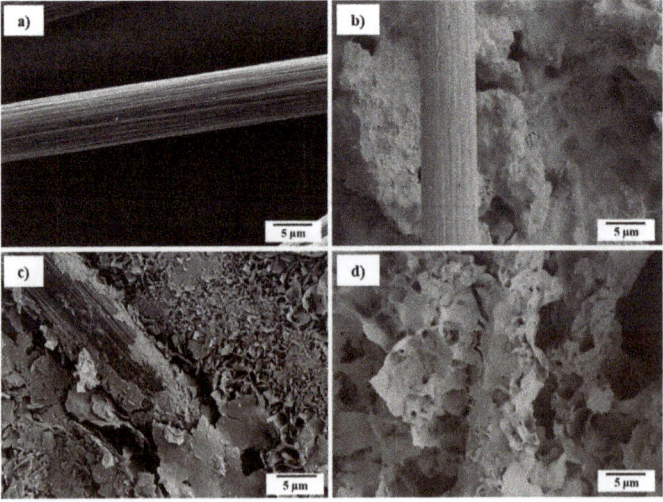

Figure 3. Scanning electron microscopy of (**a**) untreated fibers and in (**b**) cement; as well as (**c**) H_2O_2-treated fibers and (**d**) *aqua regia* and $CaCl_2$-treated fibers in cement.

Table 1. Water contact angle of chemically treated fibers and setting kinetics of calcium phosphate cement (CPC) composites.

Chemical Treatment	Contact Angle (°)	Initial Setting Time (min)	Final Setting Time (min)
Pure CPC	-	25.0 ± 0.5	105 ± 5
untreated	71 ± 4	20.0 ± 0.5	65 ± 2
H_2O_2	64 ± 6	21.0 ± 0.5	80 ± 2
aqua regia followed by $CaCl_2$	62 ± 4	18.5 ± 0.5	50 ± 2

2.3. Mechanical Properties

Figure 4 shows the bending strength (Figure 4a) and work of fracture to 5% strain (WOF5, Figure 4b) of CPC reinforced with differently modified C-fibers. Pure cement shows brittle behavior, as illustrated in SEM micrographs of the crack plane (Figure 5a) and a negligible WOF5 of 0.02 kJ·m^{-2} (Figure 4b). Unreinforced specimens are stressed until a single crack starts at a critical defect and propagates catastrophically. Only a few deviations from the direct crack path are observed, which are due to unreacted α-TCP particles or the crystallite structure of the matrix itself (Figure 5a). In the case of a C-fiber reinforcement, the addition of fibers leads to a stabilization of the crack opening and therefore steady-state cracking [29]. Composites show significantly increased strength and WOF5, even at the incorporation of only 1 wt % of untreated C-fibers. Strength was doubled, whereas the WOF5 was increased by a factor of 35 from 0.02 to 0.70 kJ·m^{-2}. This effect was already described in literature [12,18,19] and is explained by changes in the force absorption and cracking mechanism due to the incorporation of a second phase. The main effects are crack tilting and twisting, as intensively studied by Faber and Evans [30–32]. When incorporating more fibers (2 and 3 wt % C-fibers) the strength further increases, due to a higher fraction of the stronger reinforcement phase. Additionally, the WOF5 increases because the composite benefits from the higher interface ratio between fiber and matrix, leading to higher friction due to additional pull-outs.

However, untreated C-fibers show no chemical interaction with the CPC matrix. Thus, the main energy-consuming effect is related to fiber pull-out. In the case of H_2O_2-treated fibers, the slightly higher hydrophilicity, as well as the occurrence of hydroxy groups on the fiber surface, leads to a weak chemical interaction with Ca^{2+} in the cement matrix. Therefore, the force transfer from matrix to fiber is enhanced and, consequently, the bending strength increases in comparison to an untreated fiber reinforcement (Figure 4a). Beyond, weak interfacial forces increase the resistance against fiber pull-out and therefore the energy consumption, resulting in an increased WOF5 (Figure 4b).

In the case of an *aqua regia*/$CaCl_2$ pretreatment, the incorporation of fibers with polar carbonyl and carboxyl surface groups causes changes in the cracking mechanism, since these groups already bound Ca^{2+} chemically and show strong interactions with the CPC matrix. Here, the force transfer from matrix to fiber is working most effectively and, therefore, the bending strength in comparison to unmodified fibers is almost doubled even at only 1 wt % fiber content, further increasing with raised fiber fraction. SEM micrographs (Figure 5d) reveal a change in the crack pattern. In comparison to the other treatments (Figure 5b,c), a higher number of cracks with smaller widths are visible. Due to their interfacial properties bridging fibers are able to sustain the total load and transfer the force back into the matrix through interfacial shear. Consequently, this leads to the formation of another crack. This process is repeated several times, resulting in multiple cracking [33]. This force transfer also becomes evident in the crack path. Whereas for untreated and H_2O_2-treated fiber composites, cracks were propagating along the fiber–matrix interface, in CPC with *aqua regia*/$CaCl_2$-treated C-fibers, cracks were stopped, split, deflected, and propagate through the matrix material instead of only along the interface. This indicates that the resistance against pull-out and delamination is higher, and the error size distribution is augmented. Consequently, more cracks are opened. In the case of WOF5 no significant increase with fiber content is noticeable. In contrast to untreated and H_2O_2-treated fibers, the composite with *aqua regia*/$CaCl_2$ fibers does not benefit from a higher fiber content and interface ratio. However, a fiber fraction as low as 1 wt % is sufficient to initiate steady-state cracking. This mechanism is more effective in reinforcing and therefore both the strength and WOF5 are the highest after *aqua regia*/CaCl$_2$ treatment.

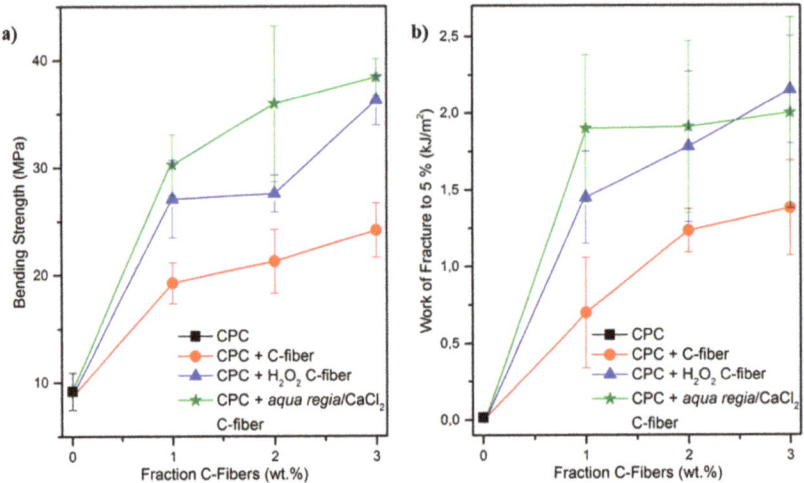

Figure 4. (a) Bending strength and (b) work of fracture up to 5% strain of C-fiber reinforced calcium phosphate cement (CPC) depending on the fiber fraction for different chemical fiber surface modifications.

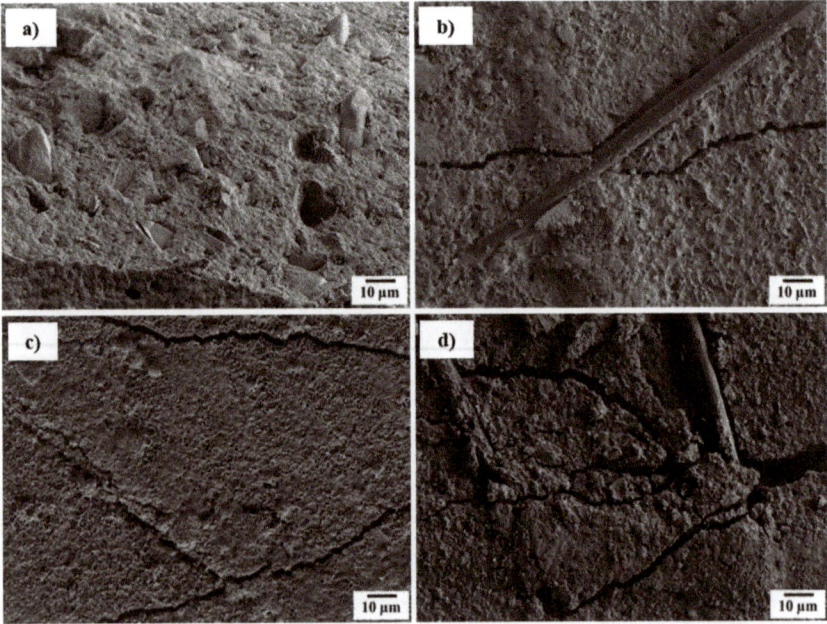

Figure 5. Scanning electron microscopy of (a) the fracture plane for pure calcium phosphate cement (CPC), and crack propagation at 1% strain for: (b) untreated; (c) H$_2$O$_2$-treated and (d) *aqua regia*- and CaCl$_2$-treated fiber-reinforced CPC with a fiber content of 1 wt %.

2.4. In Vitro Biocompatibility.

An elution test was performed to assess the biocompatibility of the cement and a possible influence of the fiber reinforcement. CPC disks were immersed and dissolved in cell culture medium, which was analyzed with regard to ionic changes and used to culture osteoblast-like cells of the cell line MG63. Cell proliferation was determined by cell detachment and counting (Figure 6a). After 3 days of incubation, the cell number was slightly higher than the seeding concentration for all samples. However, cells proliferated during culture. After 10 days, the cell number was between 60- and 70-fold, and although the reference (fresh media) had an even higher cell number (100-fold), none of the dissolution products of the samples had an inhibiting effect on osteoblast proliferation. The cell activity measured by the reaction with water-soluble tetrazolium (WST) reagent exhibited a similar behavior (Figure 6b). Here, the activity almost doubled every 3–4 days. Again, the metabolic activity of the reference was slightly higher than for the cement samples, but was steadily increasing. In both cases, no difference between the pure CDHA and the fiber-reinforced CDHA could be detected. Also, the fiber treatment had no impact on the cellular behavior. The results had a very low standard deviation indicating that the cellular response to ion release and adsorption was constant and was, therefore, reliable.

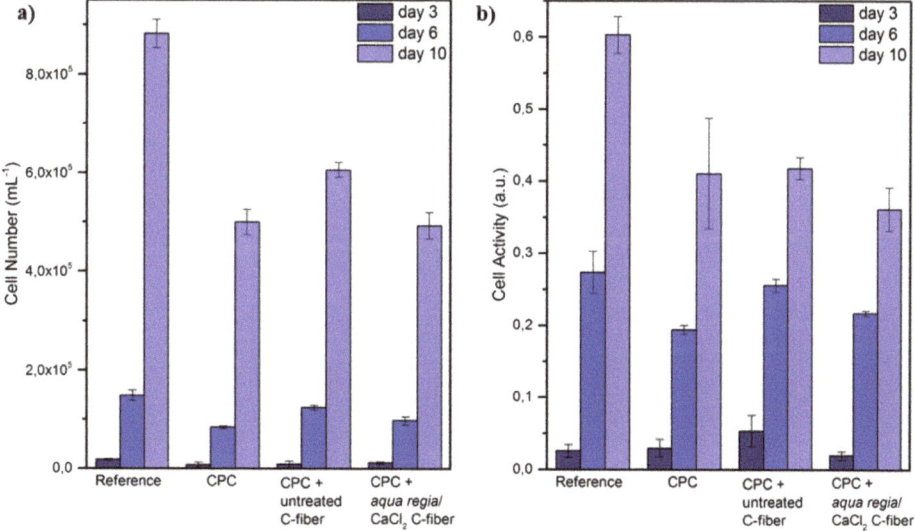

Figure 6. (**a**) Cell number and (**b**) cell activity of osteoblast-like cell line MG63. Cells were cultivated with medium from dissolution tests over 10 days.

Since changes in the ion concentration are the most influential factor on cellular behavior for calcium phosphate-based biomaterials [34,35], the supernatant of dissolving samples was investigated with inductively coupled mass spectrometry (ICP-MS). Both Ca^{2+} and PO_4^{3-} changes in the culture media were calculated against fresh media for each day (Figure 7a). Independent of fiber addition and fiber treatment, for all samples phosphate ions were released into the media, whereas calcium ions were adsorbed on the sample surface. The PO_4^{3-} release showed a maximum of around 21 mg·L^{-1} and decreased during dissolution to almost no release after 11 days. In contrast to that, the adsorption of Ca^{2+} ions started with a low adsorption and increased steadily to a maximum adsorption of around 33 mg·L^{-1} after 11 days. Additionally, the total release over 11 days was summarized, (Figure 7b). Again, there was no significant difference between pure CPC and fiber-reinforced CPC samples. The cumulative PO_4^{3-} release ranged between 100 and 130 mg·L^{-1} and the cumulative Ca^{2+}

adsorption was around 150 to 200 mg·L^{-1}. Although oxidized C-fibers showed a stronger tendency to bind Ca^{2+} ions, increased adsorption from culture media was not observed. On the contrary, over 11 days, *aqua regia*/$CaCl_2$-pretreated samples showed the lowest adsorption. Considering the pre-saturation of C-fibers in the additional $CaCl_2$ exposure step, this effect is consistent. As already known from literature [36,37], all ion concentration changes were in a non-critical range and therefore the influence on cell proliferation and activity was quite low. To overcome a reduced Ca^{2+} content, the cement could be supplemented with Sr^{2+} ions triggering similar pathways in osteoblasts [38–41]. The overall Ca^{2+} adsorption and PO_4^{3-} release is high, but the whole system around an implant must be considered. In vivo, the implant is surrounded by body fluids ensuring a steady removal of dissolution products and the supply with fresh nutrients. This would stabilize the ion concentration for cellular ingrowth.

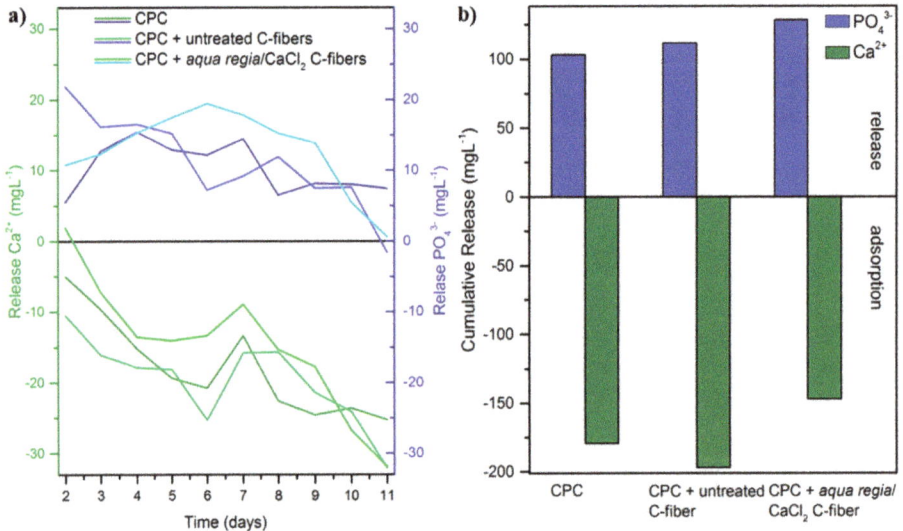

Figure 7. (a) Ca^{2+} and PO_4^{3-} release during dissolution of calcium phosphate cement (CPC) with and without fiber-reinforcement, as well as (b) cumulative release and adsorption, respectively, over a dissolution period of 11 days.

3. Materials and Methods

3.1. Fiber Modification

C-fibers with an average length of 10 mm and a diameter of 7 µm were desized in boiling propan-2-ol followed by a chemical pre-functionalization with either hydrogen peroxide (H_2O_2, 30 vol %, Carl Roth, Karlsruhe, Germany) or *aqua regia* for 40 min. Aqua regia was prepared by mixing 75 of 37 vol % hydrochloric acid (HCl, Carl Roth, Karlsruhe, Germany) and 25 of 65 vol % nitric acid (HNO_3, Carl Roth, Karlsruhe, Germany). Finally, pretreated C-fibers were stirred in 1 M $CaCl_2$ solution for 24 h at 80 °C, prepared by dissolving calcium hydride (CaH_2, Alfa Aesar, Karlsruhe, Germany) in ultra-pure water, followed by the dissolution of the resulting calcium hydroxide precipitate in HCl.

Physicochemical alterations of the fibers' surfaces were characterized by using X-ray photoelectron spectroscopy (XPS, Quantum 2000, Physical Electronics, Chanhassen, MN, USA). Single peak fittings for both spectra, O1s and C1s, were performed with the software "fityk" using Gaussian equations. Contact angle analyses (Tensiometer DCAT21, Dataphysics, Filderstadt, Germany) of modified fibers were carried out in water. For this purpose, the surface tension of water γ_{lv} was estimated by the

Wilhelmy plate method. Using the measured mass m of the advancing cycle and the fiber diameter d, which was estimated from SEM micrographs to be 6.7 µm, the contact angle θ was calculated according to Equation (1):

$$\cos \theta = \frac{\Delta m \, g}{\pi \, d \, \gamma_{lv}} \qquad (1)$$

3.2. Cement Synthesis

α-tricalcium phosphate (α-TCP) was prepared by sintering calcium hydrogen phosphate ($CaHPO_4$, purity ≥ 98%, Mallinckrodt-Baker, Griesheim, Germany) and calcium carbonate ($CaCO_3$, purity ≥ 99%, Merck, Darmstadt, Germany) in a molar ratio of 2:1 for 5 h at 1400 °C followed by quenching to room temperature. The sintered cake was crushed and passed through a 125-µm sieve followed by milling in a planetary ball mill (PM400 Retsch, Haan, Germany) at 200 rpm for 4 h [42]. A solution of 1 M trisodium citrate ($Na_3C_6H_5O_7$, Carl Roth, Karlsruhe, Deutschland) and 2.5 wt % disodium hydrogen phosphate (Na_2HPO_4, Carl Roth, Karlsruhe, Deutschland) was added to the mixture of α-TCP and 1 to 3 wt % C-fibers with a powder to liquid ratio (PLR) of 3 g·mL^{-1}. For the initial setting of CDHA, the samples were exposed to 100% humidity for 4 h at 37 °C and thereafter transferred into distilled water for the final setting period of 7 days. Mixing α-TCP powder with an aqueous solution leads to the dissolution of TCP followed by the precipitation of calcium deficient hydroxyapatite (CDHA) (Equation (2)):

$$3 \, \alpha \, Ca_3(PO_4)_2 + H_2O \rightarrow Ca_9(PO_4)_5(HPO_4)OH \qquad (2)$$

Sodium dihydrogen phosphate (Na_2HPO_4) accelerates this reaction, reducing the initial setting time for a PLR of 3 g mL^{-1} to 25 min. Additionally, sodium citrate ($Na_3C_6H_5O_7$) was added to avoid particle agglomeration and pore formation in the set cement. This CPC system was reported in detail by Gbureck et al. [43,44]. The cement was characterized by X-ray diffraction (XRD, D5000, SIEMENS Diffractometer, Berlin, Germany) using CuK α radiation (λ = 0.15405 nm), 40 kV operation voltage, and 30 mA operating current. The scanning speed during measurement was 0.02°/s for the angular range of 25–55° 2θ. Scanning electron microscopy (SEM, Sigma VP, Carl-Zeiss, Oberkochen, Germany) of pure cement, reinforced cement, and fractured specimens was performed after 7 days of setting followed by drying. The kinetics of the setting reaction were characterized using a Gilmore needle test to estimate the initial (needle diameter 2.12 mm, weight 113 g) and final (needle diameter 1.06 mm, weight 454 g) setting time. Hereby, plate specimens with a diameter of 20 mm were used.

Mechanical properties were characterized by three-point bending tests with a support span of L = 20 mm using a universal testing machine (Z020, Zwick, Ulm, Germany). The loading rate was 1 mm·min^{-1} and the preload was set to 0.1 N. The bending strength σ_b was calculated following Equation (3). The work of fracture WOF was estimated analogue to Xu et al. by using Equation (4) [19]. Hereby, the area between displacement $s_0 = 0$ mm and the displacement at failure $s_{failure}$ of the force F was calculated and normalized to the crosshead of the sample (width b, height h). However, due to the strong fiber orientation dependence on the residual strength at high strain, the error of the WOF until failure with ultimate strains up to 50% is vast. For a more expressive comparison, the WOF was calculated to the limit of 5% strain (WOF5), which is closer to the actual strain of bone [45] and therefore more meaningful for clinical applications as well. The WOF5 was calculated using only the area up to a strain of 5% (displacement s_5), as proposed in Equation (5).

$$\sigma_b = \frac{M_b}{W} = \frac{3FL}{2bh^2} \qquad (3)$$

$$WOF = \frac{\int_{s_0}^{s_{failure}} F ds}{b \times h} \qquad (4)$$

$$WOF5 = \frac{\int_{s0}^{s5} F ds}{b \times h} \tag{5}$$

3.3. In Vitro Biocompatibility

The biological behavior of fiber reinforced CPC was investigated by an elution test with the osteoblast-like cell line MG63. For this purpose, disks with a diameter of 15 mm and a height of 2 mm were used in 24-well plates. Samples ($n = 6$) were immersed in 1 mL cell culture media Dulbecco's Modified Eagle's Medium (DMEM-F12, life technologies, Carlsbad, CA, USA) supplemented with 1% penicillin/streptomycin (life technologies, Carlsbad, CA, USA) and 10% fetal calf serum (FCS, life technologies, Carlsbad, CA, USA). For dissolution, samples were stored at 37 °C and 5% CO_2. Media was exchanged daily and transferred to cell culture. MG63 was seeded with $8.333 \cdot 10^3$ cells·mL^{-1} and cultured on cell culture plastic with $n = 4$ in 48-well plates. After a cultivation of 1 day, cells were exposed to the first dissolution media extracted from the samples. The day of the first exposure was defined as day 0. Fresh media served as reference. Cells were incubated at 37 °C and 5% CO_2. Samples under investigation included CPC without fiber addition, CPC with 1% untreated C-fibers, and CPC with 1% C-fibers treated with *aqua regia* and $CaCl_2$. After 3, 6, and 10 days, cell number and cell activity were determined. The latter was quantified with a water-soluble tetrazolium (WST-1) test. For this purpose, cells were incubated at 37 °C for 30 min with WST-1 reagent. The cell activity was analyzed with a spectrometer (Tecan, Männedorf, Switzerland) at a wavelength of 450 nm. For the calculation of cell numbers, cells were first detached by incubation with Accutase (PAA Laboratories GmbH, Pasching, Austria) for 10 min at 37 °C. Subsequently detachment was stopped by media addition. Cell counting was performed on a Casy counter (Roche Innovatis AG, Bielefeld, Germany) in isotone solution.

Additionally, the supernatant was investigated in regard to the ion release from dissolving samples. Inductively coupled mass spectrometry (ICP-MS, Varian, Darmstadt, Germany) was used with standard solutions of 100, 500, and 1000 ppb for calcium and phosphorus ions. The adsorption and release of ions was calculated with respect to the ion concentration in fresh media.

4. Conclusions

The mechanical properties of CPC in terms of strength and work of fracture could be significantly enhanced by incorporating chemically modified C-fibers. The fiber activation has a strong influence on the heterogeneous nucleation of calcium phosphate crystals on the fiber surface and, consequently, on the fiber–matrix interface. It was shown that besides crack deflection and fiber pull-out in case of a weak fiber–matrix interface, multiple matrix cracking occurred preferably when activated fibers were utilized. In vitro biocompatibility tests using the osteoblast-like cell line MG63, showed no inhibitory effects of the fiber–cement composite on cell proliferation and activity. Thus, these novel CPC composites might be of particular interest for the design of injectable and load-bearing bone substitutes.

Acknowledgments: The German Research Council (DFG) is thankfully acknowledged for the financial support of the projects MU1803/16-1 and GB1/21-1. The SEM facilities of the Jena Center for Soft Matter (JCSM) were established with grants of the DFG and the European Funds for Regional Development (EFRE).

Author Contributions: A.B. performed fiber modification and characterization, and also designed and performed mechanical experiments. S.M. performed biological experiments. A.T. designed the fiber modifications. U.G. developed and prepared the cement raw powder. A.B., S.M. and F.A.M. wrote the paper. All authors significantly contributed in discussing the results.

Conflicts of Interest: The authors declare no conflict of interest.

References

1. Dorozhkin, S.V. Calcium orthophosphates: Occurrence, properties, biomineralization, pathological calcification and biomimetic applications. *Biomatter* **2011**, *1*, 121–164. [CrossRef] [PubMed]
2. Hench, L.L. Bioceramics. *J. Am. Ceram. Soc.* **1998**, *81*, 1705–1728. [CrossRef]

3. Osborn, J.F.; Newesely, H. The material science of calcium phosphate ceramics. *Biomaterials* **1980**, *1*, 108–111. [CrossRef]
4. Reznikov, N.; Shahar, R.; Weiner, S. Bone hierarchical structure in three dimensions. *Acta Biomater.* **2014**, *10*, 3815–3826. [CrossRef] [PubMed]
5. Brown, W.E.; Chow, L.C. Dental Resptorative Cement Pastes. United States Patent US-4518430 A, 21 May 1985.
6. Bohner, M. Reactivity of calcium phosphate cements. *J. Mater. Chem.* **2007**, *17*, 3980–3986. [CrossRef]
7. Bohner, M.; Gbureck, U.; Barralet, J.E. Technological issues for the development of more efficient calcium phosphate bone cements: A critical assessment. *Biomaterials* **2005**, *26*, 6423–6429. [CrossRef] [PubMed]
8. Canal, C.; Ginebra, M.P. Fibre-reinforced calcium phosphate cements: A review. *J. Mech. Behav. Biomed.* **2011**, *4*, 1658–1671. [CrossRef] [PubMed]
9. Larsson, S.; Hannink, G. Injectable bone-graft substitutes: Current products, their characteristics and indications, and new developments. *Injury* **2011**, *42* (Suppl. S2), S30–S34. [CrossRef] [PubMed]
10. Von Gonten, A.S.; Kelly, J.R.; Antonucci, J.M. Load-bearing behavior of a simulated craniofacial structure fabricated from a hydroxyapatite cement and bioresorbable fiber-mesh. *J. Mater. Sci. Mater. Med.* **2000**, *11*, 95–100. [CrossRef] [PubMed]
11. Dorozhkin, S.V. Calcium orthophosphates in nature, biology and medicine. *Materials* **2009**, *2*, 399–498. [CrossRef]
12. Krüger, R.; Groll, J. Fiber reinforced calcium phosphate cements – on the way to degradable load bearing bone substitutes? *Biomaterials* **2012**, *33*, 5887–5900. [CrossRef] [PubMed]
13. Geffers, M.; Groll, J.; Gbureck, U. Reinforcement strategies for load-bearing calcium phosphate biocements. *Materials* **2015**, *8*, 2700. [CrossRef]
14. Zorn, K.; Vorndran, E.; Gbureck, U.; Müller, F.A. Reinforcement of a magnesium-ammonium-phosphate cement with calcium phosphate whiskers. *J. Am. Ceram. Soc.* **2015**, *98*, 4028–4035. [CrossRef]
15. Nezafati, N.; Moztarzadeh, F.; Hesaraki, S.; Mozafari, M. Synergistically reinforcement of a self-setting calcium phosphate cement with bioactive glass fibers. *Ceram. Int.* **2011**, *37*, 927–934. [CrossRef]
16. Maenz, S.; Hennig, M.; Mühlstädt, M.; Kunisch, E.; Bungartz, M.; Brinkmann, O.; Bossert, J.; Kinne, R.W.; Jandt, K.D. Effects of oxygen plasma treatment on interfacial shear strength and post-peak residual strength of a plga fiber-reinforced brushite cement. *J. Mech. Behav. Biomed.* **2016**, *57*, 347–358. [CrossRef] [PubMed]
17. Dos Santos, L.A.; De Oliveira, L.C.; Da Silva Rigo, E.C.; Carrodéguas, R.G.; Boschi, A.O.; Fonseca de Arruda, A.C. Fiber reinforced calcium phosphate cement. *Artif. Organs* **2000**, *24*, 212–216. [CrossRef] [PubMed]
18. Xu, H.H.K.; Eichmiller, F.C.; Barndt, P.R. Effects of fiber length and volume fraction on the reinforcement of calcium phosphate cement. *J. Mater. Sci. Mater. Med.* **2001**, *12*, 57–65. [CrossRef] [PubMed]
19. Xu, H.H.K.; Eichmiller, F.C.; Giuseppetti, A.A. Reinforcement of a self-setting calcium phosphate cement with different fibers. *J. Biomed. Mater. Res.* **2000**, *52*, 107–114. [CrossRef]
20. National Institute of Standards and Technology. Nist X-ray Photoelectron Spectroscopy Database, version 4.1. Available online: http://srdata.nist.gov/xps/ (accessed on 24 January 2018).
21. Zielke, U.; Hüttinger, K.J.; Hoffman, W.P. Surface-oxidized carbon fibers: I. Surface structure and chemistry. *Carbon* **1996**, *34*, 983–998. [CrossRef]
22. Figueiredo, J.L.; Pereira, M.F.R.; Freitas, M.M.A.; Órfão, J.J.M. Modification of the surface chemistry of activated carbons. *Carbon* **1999**, *37*, 1379–1389. [CrossRef]
23. Zhuang, Q.L.; Kyotani, T.; Tomita, A. DRIFT and TK/TPD analyses of surface oxygen complexes formed during carbon gasification. *Energy Fuels* **1994**, *8*, 714–718. [CrossRef]
24. Zhou, J.-H.; Sui, Z.-J.; Zhu, J.; Li, P.; Chen, D.; Dai, Y.-C.; Yuan, W.-K. Characterization of surface oxygen complexes on carbon nanofibers by TPD, XPS AND FT-IR. *Carbon* **2007**, *45*, 785–796. [CrossRef]
25. Chevalier, J.; Gremillard, L. Ceramics for medical applications: A picture for the next 20 years. *J. Eur. Ceram. Soc.* **2009**, *29*, 1245–1255. [CrossRef]
26. Bohner, M.; Merkle, H.P.; Landuyt, P.V.; Trophardy, G.; Lemaitre, J. Effect of several additives and their admixtures on the physico-chemical properties of a calcium phosphate cement. *J. Mater. Sci. Mater. Med.* **2000**, *11*, 111–116. [CrossRef] [PubMed]

27. Bohner, M.; Lemaitre, J.; Ring, T.A. Effects of sulfate, pyrophosphate, and citrate ions on the physicochemical properties of cements made of β-tricalcium phosphate-phosphoric acid-water mixtures. *J. Am. Ceram. Soc.* **1996**, *79*, 1427–1434. [CrossRef]
28. Fernandez, E.; Gil, F.J.; Ginebra, M.P.; Driessens, F.C.M.; Planell, J.A.; Best, S.M. Calcium phosphate bone cements for clinical applications. Part II: Precipitate formation during setting reactions. *J. Mater. Sci. Mater. Med.* **1999**, *10*, 177–183. [CrossRef] [PubMed]
29. Marshall, D.B.; Cox, B.N.; Evans, A.G. The mechanics of matrix cracking in brittle-matrix fiber composites. *Acta Metall.* **1985**, *33*, 2013–2021. [CrossRef]
30. Evans, A.G.; Faber, K.T. Crack-growth resistance of microcracking brittle materials. *J. Am. Ceram. Soc.* **1984**, *67*, 255–260. [CrossRef]
31. Faber, K.T.; Evans, A.G. Crack deflection processes—I. Theory. *Acta Metall.* **1983**, *31*, 565–576. [CrossRef]
32. Faber, K.T.; Evans, A.G. Crack deflection processes—II. Experiment. *Acta Metall.* **1983**, *31*, 577–584. [CrossRef]
33. Li Victor, C.; Leung Christopher, K.Y. Steady-state and multiple cracking of short random fiber composites. *J. Eng. Mech.* **1992**, *118*, 2246–2264.
34. Schamel, M.; Barralet, J.E.; Groll, J.; Gbureck, U. In vitro ion adsorption and cytocompatibility of dicalcium phosphate ceramics. *Biomater. Res.* **2017**, *21*, 10. [CrossRef] [PubMed]
35. Gustavsson, J.; Ginebra, M.P.; Planell, J.; Engel, E. Osteoblast-like cellular response to dynamic changes in the ionic extracellular environment produced by calcium-deficient hydroxyapatite. *J. Mater. Sci. Mater. Med.* **2012**, *23*, 2509–2520. [CrossRef] [PubMed]
36. Meleti, Z.; Shapiro, I.M.; Adams, C.S. Inorganic phosphate induces apoptosis of osteoblast-like cells in culture. *Bone* **2000**, *27*, 359–366. [CrossRef]
37. Liu, Y.K.; Lu, Q.Z.; Pei, R.; Ji, H.J.; Zhou, G.S.; Zhao, X.L.; Tang, R.K.; Zhang, M. The effect of extracellular calcium and inorganic phosphate on the growth and osteogenic differentiation of mesenchymal stem cells in vitro: Implication for bone tissue engineering. *Biomed. Mater.* **2009**, *4*, 025004. [CrossRef] [PubMed]
38. Tarafder, S.; Dernell, W.S.; Bandyopadhyay, A.; Bose, S. SrO^- and MgO^- doped microwave sintered 3D printed tricalcium phosphate scaffolds: Mechanical properties and in vivo osteogenesis in a rabbit model. *J. Biomed. Mater. Res. Part B* **2015**, *103*, 679–690. [CrossRef] [PubMed]
39. Thormann, U.; Ray, S.; Sommer, U.; ElKhassawna, T.; Rehling, T.; Hundgeburth, M.; Henß, A.; Rohnke, M.; Janek, J.; Lips, K.S.; et al. Bone formation induced by strontium modified calcium phosphate cement in critical-size metaphyseal fracture defects in ovariectomized rats. *Biomaterials* **2013**, *34*, 8589–8598. [CrossRef] [PubMed]
40. Ni, G.X.; Chiu, K.Y.; Lu, W.W.; Wang, Y.; Zhang, Y.G.; Hao, L.B.; Li, Z.Y.; Lam, W.M.; Lu, S.B.; Luk, K.D.K. Strontium-containing hydroxyapatite bioactive bone cement in revision hip arthroplasty. *Biomaterials* **2006**, *27*, 4348–4355. [CrossRef] [PubMed]
41. Meininger, S.; Mandal, S.; Kumar, A.; Groll, J.; Basu, B.; Gbureck, U. Strength reliability and in vitro degradation of three-dimensional powder printed strontium-substituted magnesium phosphate scaffolds. *Acta Biomater.* **2016**, *31*, 401–411. [CrossRef] [PubMed]
42. Kanter, B.; Geffers, M.; Ignatius, A.; Gbureck, U. Control of in vivo mineral bone cement degradation. *Acta Biomater.* **2014**, *10*, 3279–3287. [CrossRef] [PubMed]
43. Gbureck, U.; Barralet, J.E.; Spatz, K.; Grover, L.M.; Thull, R. Ionic modification of calcium phosphate cement viscosity. Part i: Hypodermic injection and strength improvement of apatite cement. *Biomaterials* **2004**, *25*, 2187–2195. [CrossRef] [PubMed]
44. Gbureck, U.; Barralet, J.E.; Radu, L.; Klinger, H.G.; Thull, R. Amorphous α-tricalcium phosphate: Preparation and aqueous setting reaction. *J. Am. Ceram. Soc.* **2004**, *87*, 1126–1132. [CrossRef]
45. Currey, J.D. Physical characteristics affecting the tensile failure properties of compact bone. *J. Biomech.* **1990**, *23*, 837–844. [CrossRef]

© 2018 by the authors. Licensee MDPI, Basel, Switzerland. This article is an open access article distributed under the terms and conditions of the Creative Commons Attribution (CC BY) license (http://creativecommons.org/licenses/by/4.0/).

Article

Blood Vessel Formation and Bone Regeneration Potential of the Stromal Vascular Fraction Seeded on a Calcium Phosphate Scaffold in the Human Maxillary Sinus Floor Elevation Model

Elisabet Farré-Guasch [1,2], Nathalie Bravenboer [3], Marco N. Helder [2], Engelbert A. J. M. Schulten [2], Christiaan M. ten Bruggenkate [2] and Jenneke Klein-Nulend [1,*]

1. Department of Oral Cell Biology, Academic Centre for Dentistry Amsterdam (ACTA), University of Amsterdam and Vrije Universiteit Amsterdam, Amsterdam Movement Sciences, Amsterdam 1081 LA, The Netherlands; e.farreguasch@acta.nl
2. Department of Oral and Maxillofacial Surgery, VU University Medical Center/Academic Centre for Dentistry Amsterdam (ACTA), Amsterdam Movement Sciences, Amsterdam 1081 HV, The Netherlands; m.helder@vumc.nl (M.N.H.); eajm.schulten@vumc.nl (E.A.J.M.S.); cmtenbruggenkate@alrijne.nl (C.M.t.B.)
3. Department of Clinical Chemistry, VU University Medical Center, Amsterdam Movement Sciences, Amsterdam 1007 MB, The Netherlands; n.bravenboer@vumc.nl
* Correspondence: j.kleinnulend@acta.nl; Tel.: +31-(0)-205980-881; Fax: +31-(0)-205-980-333

Received: 15 December 2017; Accepted: 18 January 2018; Published: 20 January 2018

Abstract: Bone substitutes are used as alternatives for autologous bone grafts in patients undergoing maxillary sinus floor elevation (MSFE) for dental implant placement. However, bone substitutes lack osteoinductive and angiogenic potential. Addition of adipose stem cells (ASCs) may stimulate osteogenesis and osteoinduction, as well as angiogenesis. We aimed to evaluate the vascularization in relation to bone formation potential of the ASC-containing stromal vascular fraction (SVF) of adipose tissue, seeded on two types of calcium phosphate carriers, within the human MSFE model, in a phase I study. Autologous SVF was obtained from ten patients and seeded on β-tricalcium phosphate (n = 5) or biphasic calcium phosphate carriers (n = 5), and used for MSFE in a one-step surgical procedure. After six months, biopsies were obtained during dental implant placement, and the quantification of the number of blood vessels was performed using histomorphometric analysis and immunohistochemical stainings for blood vessel markers, i.e., CD34 and alpha-smooth muscle actin. Bone percentages seemed to correlate with blood vessel formation and were higher in study versus control biopsies in the cranial area, in particular in β-tricalcium phosphate-treated patients. This study shows the safety, feasibility, and efficiency of the use of ASCs in the human MSFE, and indicates a pro-angiogenic effect of SVF.

Keywords: angiogenesis; bone; adipose stem cells; calcium phosphates; bone substitutes; clinical translation; clinical trials; adipose

---

1. Introduction

Retention of removable dentures is often a problem for edentulous patients with severe maxillary or mandibular atrophy, a common problem in the aged population. Installation of dental implants can help these patients by providing better support for and retention of their dental prosthesis. Placing dental implants requires sufficient jaw bone volume, which is often not available, especially in the edentulous distal maxillary area of severely atrophied maxillae. In these cases, the bone volume must be increased by augmentation prior to dental implant placement.

Maxillary sinus floor elevation (MSFE) or augmentation has been introduced to dentistry in the mid-1970s (modified by Tatum in 1986) [1] and has become part of pre-prosthetic surgery. In this surgical procedure, the distal area of the maxilla is augmented by transplanting bone or bone substitutes to the bottom of the maxillary sinus. The maxillary sinus floor elevation model is unique by allowing histological examination of biopsies obtained during the preparation for dental implant placement using a hollow trephine bur.

In dentistry, autologous bone grafting is still the "gold standard" for bone augmentation. Autografting, however, has an important disadvantage, i.e., trabecular bone needs to be harvested from elsewhere in the skeleton, usually from the lateral mandible, the chin, or the iliac crest, and applied to the jaw defect. This means another operation site, including the risk of hospitalization, causing donor site morbidity, and potential complications, such as post-operative infections, and functional defects [2]. Therefore, alternative bone substitute materials have been evaluated. Calcium phosphates, such as hydroxyapatite (HA), β-tricalcium phosphate (β-TCP), and a combination of HA/β-TCP, are often used since they do not evoke adverse cellular reactions and, in time, the material is either replaced by bone or integrated into the body, depending on the degradation properties [3–5]. However, these materials have limitations due to the slow speed of osteoconduction when compared to the properties of autologous bone [6–8]. The time required is strongly determined by the time that osteogenic cells take to grow into the augmentation space and would potentially be shortened if bone regeneration-competent cells (mesenchymal stem cells; MSCs) are mixed with the bone substitute. The calcium phosphate carrier has controlled 3D properties allowing immediate colonization by MSCs of the entire volume of the ceramics, and in depth revascularization [9,10].

Endothelial cells are lining blood vessels and allow the formation of new blood capillaries by the sprouting of an existing small vessel, a phenomenon called angiogenesis. This process allows tissue growth and repair by extending and remodeling the network of blood vessels [11,12]. In contrast to small blood vessels, composed of endothelial cells surrounded by a basal lamina and loosely covered by single pericytes, larger vessels are coated with multiple layers of smooth-muscle cells and elastic and collagenous fibers [6]. These are composed by the endothelium, a thin layer of endothelial cells, separated from the surrounding outer layers by a basal lamina. The amounts of connective tissue and smooth muscle in the vessel wall vary according to the vessel's diameter and function, but the endothelial lining is always present. Pericytes have been associated mainly with stabilization and hemodynamic processes of blood vessels. Their functions are, however, much more diverse than traditionally thought. They can sense angiogenic stimuli, guide sprouting tubes, elicit endothelial survival functions, and even exhibit macrophage-like activities which make them crucial cells in the process of tissue repair and remodeling [13].

Poor angiogenesis is a common and vital barrier to tissue regeneration. Regenerating tissue over 200 µm exceeds the capacity of nutrient supply and waste removal from the tissue and, therefore, requires an intimate supply of vascular networks [14]. This has led to the use of angiogenic growth factors and/or transplantation of proangiogenic cells, such as endothelial progenitor cells (EPCs), in combination with scaffolds. The use of angiogenic growth factors and/or transplantation of these proangiogenic cells only, however, also has disadvantages, since perivascular cells, including mural cells, are obligatory for the formation of native, multilayered mature microvessels [15]. The potential of MSCs to stimulate angiogenesis holds interesting promises to the field of tissue engineering.

Adipose tissue may be easily obtained from patients using liposuction or oral surgical procedures [16]. Adipose tissue represents a promising source of MSCs, as liposuction can be performed with minimal patient discomfort and yields higher numbers of MSCs than bone marrow, which could avoid costly cell expansion to obtain a number of cells high enough for clinical use [17–19]. Risks associated with cell culturing such as pathogen contamination, spontaneous transformation and loss of proliferation and differentiation potential are thus minimized [20–22]. The stromal vascular fraction (SVF), obtained immediately after digestion of adipose tissue with collagenase and centrifugation to separate the floating adipocytes, is highly heterogeneous and contains many cell subsets, including native adipose stem cells (ASCs), mature endothelial cells, and haematopoietic cells [19]. SVF also contains macrophages, which secrete a multitude of vascular growth factors and cytokines [23]. The ASCs in SVF have been shown to attach, proliferate, and osteogenically differentiate on calcium phosphate scaffolds [24], and secrete a high number of growth factors [25]. ASCs have not only shown osteogenic potential in vivo [26,27], but also demonstrated angiogenic potential crucial for bone tissue engineering applications in mice [28]. This supports in vitro observations that ASCs in SVF secrete a variety of angiogenic and anti-apoptotic growth factors [29], and that SVF is highly enriched with CD34+CD45− cells. The CD34+ cells are capable of stimulating angiogenesis, and are involved in neovascularization processes that facilitate healing of ischemic tissues in mouse models [30]. However, whether SVF/ASCs are also effective in stimulating vascularization in humans has not been unequivocally shown so far.

In an earlier performed clinical phase I study [27], employing our one-step surgical procedure [18] in patients undergoing maxillary sinus floor elevation, we isolated SVF, and re-implanted them intraoperatively into the patient again. This study successfully showed feasibility, safety, and potential efficacy of using bioactive implants consisting of calcium phosphate carriers seeded with freshly isolated SVF containing ASCs. Since we hypothesized that the SVF will positively contribute not only to bone formation, but likely also to vascularization, the current study aimed to evaluate vascularization in relation to bone formation potential of SVF in biopsies from the previous study obtained after six months (when dental implant placement occurred). For this evaluation, a recently developed immunohistochemical staining technique for methyl methacrylate (MMA) polymer resin embedded bone biopsies [31] was applied.

2. Materials and Methods

2.1. Clinical Study Outline

This angiogenic study is an extension of a phase I/IIa clinical trial study we reported on before [27], in which safety and potential efficacy of maxillary sinus floor bone augmentation using calcium phosphate bone substitutes and a freshly-isolated adipose stem cell preparation, termed the stromal vascular fraction (SVF), was evaluated in a one-step surgical procedure in 10 partially-edentulous patients requiring dental implants for prosthetic rehabilitation. The clinical study was registered in the Netherlands Trial Registry (NTR4408), and complied with the principles of the Declaration of Helsinki of 1975, revised in 2008. All protocols were approved by the medical ethics committee (IRB) of the VU University Medical Center Amsterdam, as well as the central committee on research involving human subjects (CCMO, The Hague, The Netherlands; Dossier number: NL29581.000.09; EudraCT-number: 2009-015562-62). All patients signed a written informed consent before participation in the study. For detailed inclusion and exclusion criteria, procedures and assessments according to the study protocol of our previously performed phase I trial, one is referred to our previous report [27]. The demographic data of our patient group is summarized in Table 1.

A graphic visualization of the surgical protocol is depicted in Figure 1. Briefly, the surgery started by collecting >125 mL of adipose tissue using a syringe-based lipoaspiration. The lipoaspirate (Figure 1a) was subsequently transported to a special stem cell laboratory within the VU University Medical Center Amsterdam operation complex. Within the stem cell laboratory, the adipose tissue was processed with the CE marked Celution device (Cytori Therapeutics, Inc., San Diego, CA, USA) (Figure 1b) to obtain SVF. Viability and cell number was determined in triplicate with a Nucleocounter NC-100 (ChemoMetec A/S, Allerød, Denmark) according to the manufacturer's protocol. The release criterion was set at ≥70% viability. For implantation cells were seeded in Ringer's lactate solution in a concentration of 10^7 nucleated SVF cells ($\pm 2 \times 10^5$ ASC-like cells)/g calcium phosphate carrier (Figure 1c). Calcium phosphate carriers consisted of 100% Ceros® β-TCP with 60% porosity and granule size of 0.7–1.4 mm (Thommen Medical, Grenchen, Switzerland) or Straumann®Bone Ceramic biphasic calcium phosphate (BCP), consisting of 60% hydroxyapatite (HA) and 40% β-tricalcium phosphate (β-TCP) with 90% porosity and granule size of 0.5–1.0 mm (Straumann AG, Basel, Switzerland). After allowing attachment of the cells for 30 min and subsequent washing with Ringer's lactate solution to remove unattached cells, the carriers were implanted using a standard maxillary sinus floor elevation procedure according to the lateral "top hinge trap door" procedure of Tatum [1] (Figure 1d–f). In the case of a bilateral "split-mouth" design treatment, one side was implanted with the cell-seeded carrier, while the other (control) side was implanted with carriers undergoing the same seeding procedure, but with vehicle (Ringer's lactate solution) only.

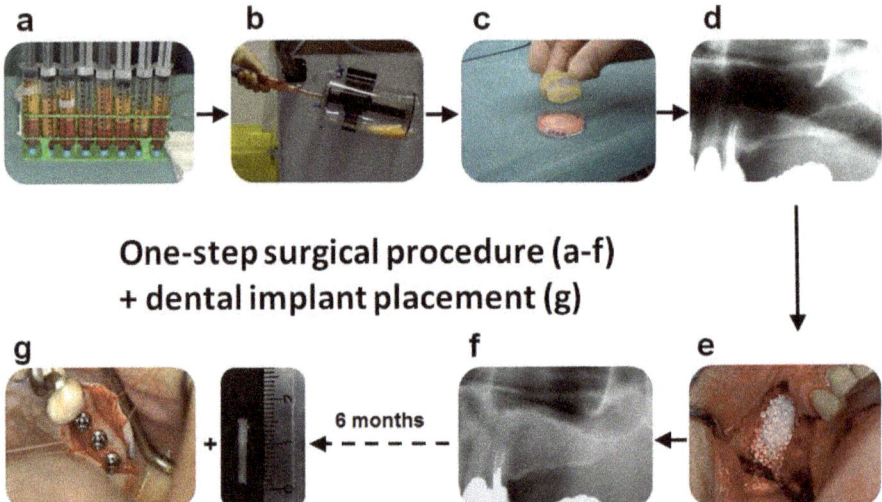

Figure 1. Concept of a sinus floor elevation with freshly isolated adipose-derived stem cells in a one-step surgical procedure. (**a**) The adipose tissue and liposuction fluid obtained by liposuction is collected in syringes; (**b**) the filled syringes are transferred into a Celution 800/CRS system to obtain the fresh stromal vascular fraction containing the adipose stem cells; and (**c**) the freshly isolated adipose stem cells are seeded onto the calcium phosphate scaffold. Unattached cells are washed off. During the short attachment period of the cells (30 min), the patient is prepared for the maxillary sinus floor elevation procedure via a lateral approach; (**d**) the anatomical selection criterion was a pre-existing (native) alveolar bone height of >4 mm obtained from the preoperative panoramic radiograph; and (**e**) after reflection of the mucoperiosteal flap, a bony window is created in the lateral wall of the maxillary sinus and carefully moved and rotated medially toward the maxillary sinus, after dissection of the maxillary sinus mucosa (trap-door technique). The calcium phosphate scaffold is inserted immediately into the patient, and the space created is filled with the bone substitute combined with the adipose stem cells. Finally, the wound is closed; (**f**) after a healing period of five months post-MSFE (prior to dental implant placement), a panoramic radiograph is made to determine the increase in vertical height bone + bone substitute at the planned dental implant positions; and (**g**) after six months, bone biopsies are taken by using a hollow burr for histomorphometrical and immunohistochemical analysis, and dental implants are placed.

Bone biopsies were obtained during dental implant surgery after a six month healing period, using hollow trephine drills with an external diameter of 3.5 mm (Straumann AG, Basel, Switzerland) under local anesthesia prior to dental implant placements (Figure 1g). After a three month osseointegration period the suprastructures were manufactured, and placed by the patient's dentist.

2.2. Biopsy Processing and Evaluation

The biopsies were fixed in 4% formaldehyde solution (Klinipath BV, Duiven, The Netherlands) at 4°C for 24 h, removed from the drill, transferred to 70% ethanol, and stored until use for histomorphometrical analysis (Figure 2), as described below. For a valid, uniform comparison of the biopsies taken from the sides that were treated with calcium phosphate only and the biopsies taken from the sides augmented with calcium phosphate and ASCs, a selection was made by two independent experienced observers. The biopsies taken from implant sides outside the augmented maxillary sinus (mainly implant position 14 and/or 24) were excluded from analysis. Per patient, one biopsy from each side was selected in the middle of the grafted area to exclude the effect of surrounding bone containing mechanically-loaded dental elements and bone near the nasal wall of the maxillary sinus. Using these selection criteria, a total of 16 selected biopsies (six from the control side, 10 from the study side) from 10 patients were studied to analyze the blood vessel formation and bone formation (Table 1, bold numbers). In addition, one sample from a transilical biopsy was used as control for the immunohistochemical analysis.

2.3. Histology and Histomorphometry

After dehydration in descending alcohol series, the bone specimens were embedded without prior decalcification in low temperature polymerizing methylmethacrylate (MMA, Merck Schuchardt OHG, Hohenbrunn, Germany). Longitudinal sections of 5 µm thickness were prepared using a Jung K microtome (R. Jung, Heidelberg, Germany). Midsagittal histological sections of each biopsy were stained with Goldner's Trichome method [6], in order to distinguish mineralized bone tissue (green) and unmineralized osteoid (red). The histological sections were divided into regions of interest (ROI) of 1 mm^2 for blinded histomorphometrical analysis, as previously described [32]. Depending on the length of the biopsy, the number of ROIs ranged from 9–15. The digital images of the scanned biopsies were analyzed, starting from the caudal side of the biopsy, and continuing towards the cranial side. This method allowed to compare similar ROIs for all biopsies (with and without stem cells) with respect to the bone regeneration and blood vessel formation in the augmented maxillary sinus. For each ROI, the bone volume (BV) was calculated as a percentage of the total tissue volume (TV), as previously described [33]. This analysis was performed by two independent blinded observations.

For each separate area of interest, the histomorphometrical measurements were performed with a computer using an electronic stage table and a Leica DC 200 digital camera (Leica, Wetzlar, Germany). The computer software used was Leica QWin© (Leica Microsystems Image Solutions, Rijswijk, The Netherlands). Digital images of the sections were acquired at 100× magnification. Consecutive ROI of 1 mm^2 each were defined and numbered throughout the whole biopsy. The transition zone (TZ) indicates the first ROI where graft material was observed when analyzing from the caudal to the cranial side of the biopsy. Because the biopsies analyzed had different lengths, we decided to define them in three regions after the transition zone (TZ) between the native bone or caudal area and the scaffold area towards cranial. The first two ROIs on the right of the transition zone were defined as region I, the two or three ROIs in the middle (even or odd numbers) as region II, and the two rightmost ROIs as region III (Figure 2).

Data from the residual native bone part of the biopsy next to the transition zone and for each area from the sinus floor towards the cranial side of the biopsy was analyzed separately (Figure 2). Similar to this method we have already been able to compare similar areas of interest for the two sides (control and test side) with respect to the bone regeneration performance indicated by the amount of osteoid and bone formed, and volume of remaining graft material [32].

Blood vessel numbers, taking into account the blood vessel size, were determined as mean value of two separate blinded counts. Blood vessel size was calculated as the total blood vessel area expressed in µm^2. According to their size, blood vessels were divided into small (0–400 µm^2) or large vessels (>400 µm^2).

2.4. Immunohistochemistry

To quantify microvessel density, immunohistochemical staining for CD34, a marker of endothelial cells, as well as stem cells, such as endothelial progenitor stem cells and hematopoietic stem cells, was performed [34,35]. The expression of smooth muscle actin (SMA), a marker of smooth muscle cells as well as pericytes, was also analyzed by immunohistochemistry [35,36]. The presence of pericytes and smooth muscle cells surrounding blood vessels has been described as a structural parameter indicative of vascular maturity.

The MMA in the sections was removed by immersing them in xylene/chloroform (Merck, Darmstadt, Germany) for 30 min at room temperature followed by two rinses with xylene. Bone sections were rehydrated in graded alcohol solutions. To block endogenous peroxidase when using the avidin biotin peroxidase complex the hydrated specimens were transferred to 3% (v/v) hydrogen peroxide in methanol for 15 min. In a set of pilot experiments, it was deduced that optimal CD34 staining results were achieved with 5 µg/mL proteinase K pretreatment for 10 min (Invitrogen, Carlsbad, CA, USA), while SMA detection was best without proteinase K predigestion. Non-specific binding of immunoglobulin G was blocked by incubation with 5% (v/v) normal serum (as appropriate for each antibody) for 60 min with 0.1% bovine serum albumin (Sigma, St Louis, MI, USA). The sections were then incubated with the primary antibodies at optimal dilutions for 2 h at room temperature (SMA Monoclonal Mouse Anti-Human Smooth Muscle Actin Clone 1A4 1:50, Dako, Carpinteria, MI, USA) and overnight incubation at 4 °C in a humidified chamber (CD34 Monoclonal Mouse Anti-Human CD34 Class II Clone QBEnd-10 1:20, Dako). Sections were rinsed in phosphate-buffered saline and incubated at room temperature with an horseradish peroxidase (HRP)-labelled polymer conjugated with secondary antibody (Envision Kit, Dako, Santa Clara, CA, USA), for 0.5 h at room temperature before detection using aminoethyl carbazole (AEC) (Invitrogen, Carlsbad, CA, USA) staining as recommended by the manufacturer's protocol. Sections were counterstained with haematoxylin (Merck, Schuchardt OHG, Hohenbrunn, Germany). Sclerostin (hSOS, Dako, Santa Clara, CA, USA) was used as positive control. Negative controls (without primary or secondary antibodies, or both) were also performed for all antibodies tested.

The determination of the number of blood vessels was performed as previously described [37]. The number of blood vessels was expressed per area of soft connective tissue (mm^2) [37]. Blood vessel size was calculated as described above.

2.5. Statistics

Data are presented as mean ± standard deviation (SD). Data analysis and statistical analysis were performed using GraphPad Prism 5 software (GraphPad Software, La Jolla, CA, USA) and IBM SPSS 23 statistical software (CircleCI, San Francisco, CA, USA). Selected biopsies from all treated patients (with vs. without stem cells) were compared between the β-TCP and BCP groups.

A paired Wilcoxon signed rank test was performed to assess whether bone volume and blood vessel number were higher at the study sides compared to control sides for each material. An unpaired nonparametric Mann Whitney U test was performed to test differences between β-TCP and BCP in control and study samples. Statistical significance was considered if p-values were <0.05.

3. Results

3.1. Clinical Evaluation and Implant Survival

Patient data and selected biopsies are listed in Table 1. As control for the immunohistochemical analysis, transiliacal bone biopsies and paraffin-embedded maxillary bone biopsies were used for the assays.

Table 1. Patient demographics (as published earlier in [27]) showing gender and age (years), body mass index, whether the patients were treated unilaterally or bilaterally ("split-mouth design"), type of calcium phosphate bone substitute used for the maxillary sinus floor elevation procedure, and the dental implant sites according to the Fédération Dentaire Internationale (FDI) system. Selected biopsies are displayed in bold. β-TCP, β-tricalcium phosphate; BCP, biphasic calcium phosphate.

Patient Number	Gender, Age (Years)	Body Mass Index	Unilateral/ Bilateral	Control/ Study Side	Graft Material	Dental Implant Positions
1	♀, 58	24.5	bilateral	Control study	β-TCP	14,15,16 24,25,26
2	♀, 46	30.4	bilateral	Study control	β-TCP	14,15,16 24,25,26
3	♂, 69	35.8	bilateral	control study	β-TCP	14,15,16 25,26,27
4	♀, 59	31.2	unilateral	study	β-TCP	24,25,26
5	♂, 64	24.8	unilateral	study	β-TCP	15,16
6	♀, 57	24.2	bilateral	study control	BCP	14,15,16 24,26
7	♂, 56	28.7	bilateral	study control	BCP	15,16,17 25,26,27
8	♀, 51	30.9	unilateral	study	BCP	14,15,16
9	♂, 52	32.1	unilateral	study	BCP	23,25,26
10	♀, 51	30.3	bilateral	study control	BCP	15,16 25,26

3.2. Data Analysis

The biopsies analyzed were obtained from a clinical phase I study in 10 patients, which was primarily aimed at assessing feasibility and safety of a one-step surgical concept applying the freshly isolated SVF containing adipose stem-cell like cells in combination with calcium phosphate bone substitutes. We are aware that our sample size is relatively low and, therefore, we mainly focused on the patterns obtained in this results section. Nevertheless, we felt that it is relevant to include the statistical analyses performed as additional information for the reader to judge our findings, but would like to state that our statistical analyses as described below should be seen as indications and not as solid proof for our statements.

3.3. Quantitative Histomorphometric Evaluation

All bone biopsies were analyzed from caudal (native bone) to cranial (scaffold area). In contrast to the previous report [27], we now divided the graft area in three distinct regions, and analyzed angiogenesis in the biopsies.

In the vehicle-treated (control) β-TCP group, we observed that the mineralized bone volume decreased from the residual native bone next to the transition zone towards the cranial side of the biopsies, and that no bone formation was present in the most cranial area, designated as region III (Figure 2a). In confirmation with the previous study, when analyzing the bone biopsies in the SVF-treated β-TCP group, higher bone volumes per total volume were found in region III in 80% (four out of five) of the patients (Figure 2b). When comparing the biopsies from the SVF-supplemented sides with the control sides, a marked although non-significant difference in the bone volumes in region III (SVF-treated vs. control: $17.7 \pm 10.3\%$ vs. $0 \pm 0\%$) ($p = 0.1$) was observed (Figure 2c). When analyzing blood vessel formation in a similar way, a higher blood vessel number was found in all regions of the scaffold area in the SVF-treated group vs. the control group (Figure 2d), but these differences did not reach significance.

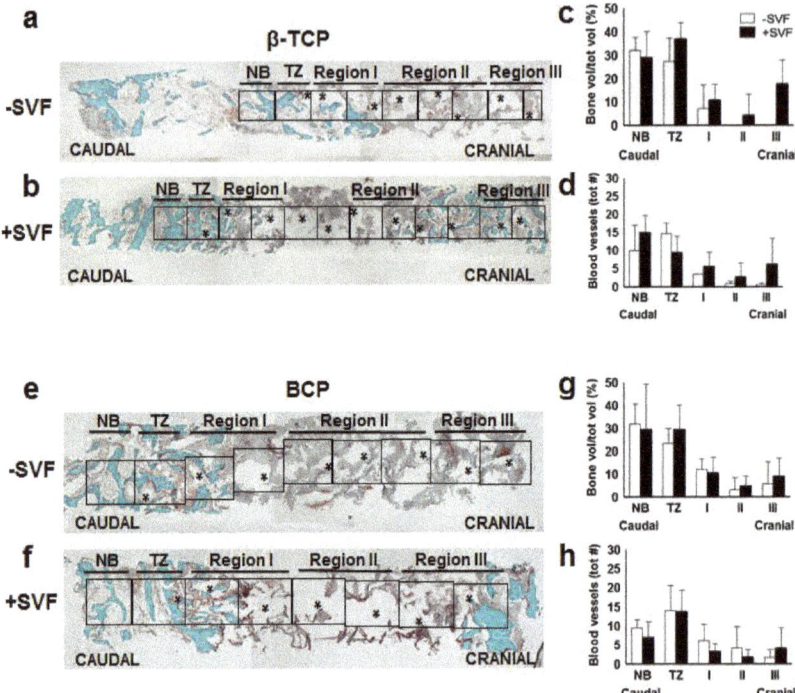

Figure 2. Histomorphometrical analysis of maxillary bone biopsies from patients treated with β-TCP (**a–d**) and BCP (**e–h**). Graphs a and b on the one side, and e and f on the other side represent typical bilateral biopsies from one and the same patient, while graphs c, d, g, and H represent quantifications of 3–5 samples. Midsagittal histological sections of each biopsy were stained with Goldner's Trichome method [3], to distinct mineralized bone tissue (green) and unmineralized osteoid (red). Biopsies were divided in consecutive 1 mm^2 regions of interest (ROIs). The transition zone (TZ) indicates the first ROI where graft material (*) was observed when analyzing from the caudal to the cranial side of the biopsy. Bone ingrowth is determined from the sinus floor towards the cranial side of the biopsies. Original magnification is 100×. Since the biopsies had different lengths we decided to define them in three regions after the transition zone. The ROI of the native bone (NB) next to the TZ was also analyzed. Percentage mineralized bone vol/tot vol from control sides without stem cells (white bars; $n = 3$), and study sides with stem cells (black bars; $n = 5$) from patients treated with β-TCP (**b**) or BCP (**f**) are depicted. Total number of blood vessels from control sides without stem cells (white bars; $n = 3$), and study sides with stem cells (black bars; $n = 5$) from patients treated with β-TCP (**d**) or BCP (**h**). β-tricalcium phosphate; BCP, biphasic calcium phosphate; NB, native bone; TZ, transition zone; SVF, stromal vascular fraction; bone vol/tot vol, bone volume/total volume; tot #, total number.

In the BCP group, the mineralized bone volume decreased towards the cranial side of the biopsies. In 33% (one out of three) of the control sides, bone formation in region III was observed (Figure 2e), whereas this occurred in 60% (three out of five) of the SVF-supplemented sides (Figure 2f). Moreover, a higher percentage of bone volume per total volume was found in SVF-treated vs. control sides in this region (BCP: 8.8 ± 8% vs. 5.6 ± 9.8%), although this was less prominent compared to the findings in the β-TCP treatments (Figure 2g). This was also the case for the blood vessel formation in region III (Figure 2h).

Comparison of the bone contents between the control and SVF-supplemented group per biomaterial (Figure 2c,g) in each region showed no significant differences. When comparing the two biomaterials for their effect on bone formation in the presence or absence of SVF, it could be concluded that β-TCP, but not BCP, displayed increased bone formation in combination with SVF, but only in region III ($p = 0.047$). In the absence of SVF, no difference between the two biomaterials was observed.

When analyzing the relative percentages of small and large blood vessels in the different regions of the scaffold area, we observed that the control sides in the β-TCP group showed a similar percentage of small and large blood vessels (~70%) in region I which both decreased gradually from caudal towards cranial to ~10% in region III (Table 2). Interestingly, the SVF-supplemented group showed a different pattern: instead of a steady decrease, a biphasic pattern could be observed, i.e., high vessel percentages (~40% with an approximate 1:1 ratio between small and large vessels) in region I, four-fold lower percentages in region II for both vessel sizes, and another ~40% and 1:1 ratio in region III.

Table 2. Percentages of small and large blood vessels in the scaffold area subregions (regions I, II, and III) from maxillary bone biopsies obtained from patients treated with β-TCP and BCP without stem cells (−SVF, $n = 3$) and with stem cells (+SVF, $n = 5$) by histomorphometrical and immunohistochemical analysis. β-TCP, β-tricalcium phosphate; BCP, biphasic calcium phosphate; SVF, stromal vascular fraction.

	Size of blood vessels	Region I	Region II	Region III	Total	Region I	Region II	Region III	Total
		\multicolumn{9}{c}{% of Total Blood Vessels}								
		\multicolumn{4}{c}{−SVF}	\multicolumn{4}{c}{+SVF}						
β-TCP	Small	68	26	6	100	40	13	47	100
	Large	69	20	11	100	47	10	43	100
BCP	Small	43	37	20	100	29	43	28	100
	Large	67	18	15	100	59	7	34	100
		\multicolumn{9}{c}{% of Total CD34+ Blood Vessels}								
	Size of blood vessels	Region I	Region II	Region III	Total	Region I	Region II	Region III	Total
		\multicolumn{4}{c}{−SVF}	\multicolumn{4}{c}{+SVF}						
β-TCP	Small	70	19	11	100	50	25	25	100
	Large	76	24	0	100	40	0	60	100
BCP	Small	70	14	16	100	30	51	19	100
	Large	69	0	31	100	18	42	40	100
		\multicolumn{9}{c}{% of Total SMA+ Blood Vessels}								
	Size of blood vessels	Region I	Region II	Region III	Total	Region I	Region II	Region III	Total
		\multicolumn{4}{c}{−SVF}	\multicolumn{4}{c}{+SVF}						
β-TCP	Small	52	48	0	100	30	24	46	100
	Large	41	46	13	100	10	20	70	100
BCP	Small	80	0	20	100	35	43	22	100
	Large	75	0	25	100	22	19	59	100

The control sides of the BCP group displayed a similar pattern as described for the control β-TCP sides, with a gradual decrease of blood vessel formation from region I to region III, although region I showed a slightly different percent ratio (65:45) between large and small vessels, which leveled to ~15% each at a 1:1 ratio (Table 2). Again, different patterns were observed in the SVF-supplemented BCP group, i.e., for the small vessels, region I and III showed equal percentages and even an increase in region II, while the large vessel counts showed the highest number in region I, low numbers in region II, and intermediate numbers in region III. Please refer to Supplementary Materials Table S1 for more detailed information on the percentages of small and large blood vessels from maxillary bone biopsies obtained from patients treated with β-TCP and BCP without stem cells and with stem cells, as analyzed by histomorphometrical and immunohistochemical analyses.

3.4. Immunohistochemistry

We found that tissue processing techniques preserved original morphology, while scratch artifacts, folds, and distensions were absent in all sections.

3.4.1. CD34+ Blood Vessels

We observed CD34 staining around vessels in the maxillary bone samples of methyl-methacrylate embedded tissue of patients augmented with β-TCP or BCP (Figure 3a–f).

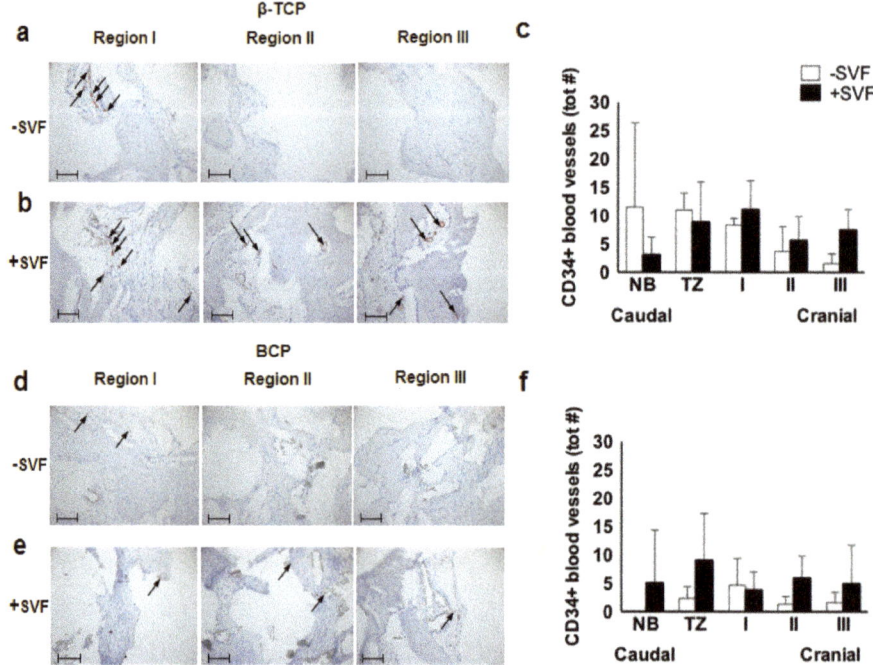

Figure 3. Immunohistochemical analysis of CD34, a marker of endothelial cells as well as stem cells such as endothelial progenitor stem cells and hematopoietic stem cells, of a maxillary bone biopsy from a patient treated with β-TCP (**a–c**) and BCP (**d–f**). Magnification: 200×. The scale bar represents 100 μm. The total number of CD34+ blood vessels of selected bone biopsies taken from control sides without stem cells (white bars; $n = 3$), and study sides with stem cells (black bars; $n = 4$) from patients treated with β-TCP (**b**) or BCP (**e**). β-TCP, β-tricalcium phosphate; BCP, biphasic calcium phosphate; NB, native bone; TZ, transition zone; SVF, stromal vascular fraction; tot #, total number.

In the control sides of the β-TCP group, the total number of CD34+ blood vessels decreased from region I to III, resulting in only scarce expression in region III (Figure 3a). More staining, particularly in regions II and III, was observed in the SVF-treated sides of the β-TCP group (Figure 3b). When quantifying the stained vessels in the β-TCP group, all three graft regions showed higher numbers of CD34+ vessels in the SVF-supplemented vs. the control sides with the largest difference in region III, but none reached statistical significance (Figure 3c).

In the BCP group, both the control (Figure 3d) and the SVF-supplemented sides (Figure 3e) showed similar patterns as described for the β-TCP group. Quantification indicated stable total CD34+ vessel numbers for the SVF-seeded scaffolds which were higher in region II and in region III when compared to the control sides (Figure 3f).

When analyzing the relative percentages of CD34+ small and large blood vessels in the different regions of the scaffold area (Table 2), we observed that in the control sides of the β-TCP group a similar percentage of small and large blood vessels was found in region I (~70% of the total CD34+ vessels in the grafted area), which decreased gradually from caudal towards cranial. In the stem cell-treated β-TCP group a comparable pattern was found except for region III, where the percentage of large blood vessels increased again to even higher values than in region I (from 40% to 60%).

In the control sides of the BCP group, the percentage of small vessels declined as described for their counterparts in the β-TCP group, but the number of large vessels again increased from region II to region III. In the SVF-supplemented augmentations comparable patterns were observed, but in region II both the small and large vessels were considerably higher, indicating more advanced angiogenesis in the core of the grafted material (Table 2).

3.4.2. SMA+ Blood Vessels

Immunostaining for SMA, indicative for more mature blood vessels, resulted in relatively low numbers of SMA+ blood vessels in the grafted area. When comparing the control (Figure 4a) and SVF-supplemented (Figure 4b) sides of the β-TCP group, the number of SMA+ vessels was low for both treatments in regions I and II. Quantification showed differential numbers of SMA+ stained vessels in region III: no SMA+ vessels in the control sides, while slightly increased SMA+ blood vessel numbers (relative to regions I and II) in the SVF-supplemented sides (Figure 4c). Similar patterns were observed in the BCP group (Figure 4d–f).

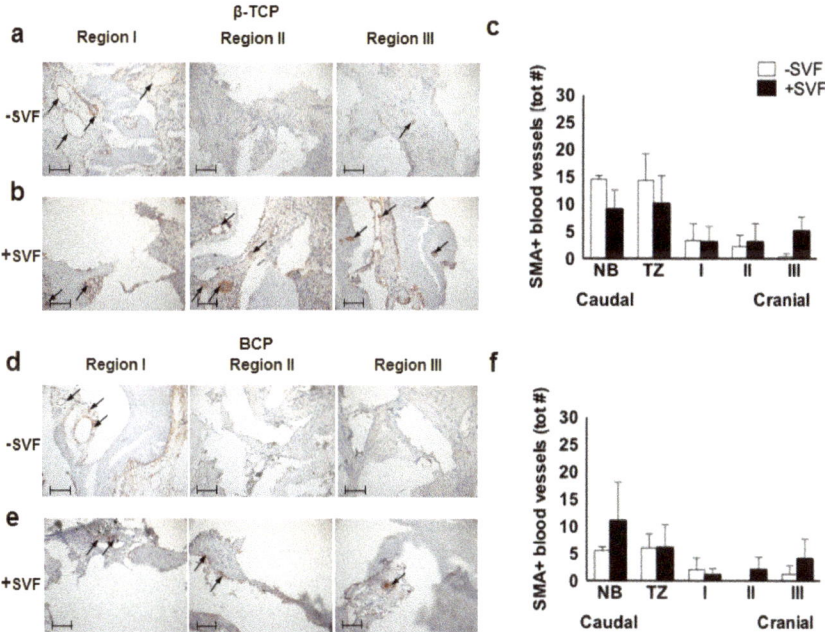

Figure 4. Immunohistochemical analysis of α-smooth muscle actin (SMA), a marker of pericytes, as well as smooth muscle cells present in the blood vessel walls, of a maxillary bone biopsy from a patient treated with β-TCP (**a–c**) and BCP (**d–f**). Magnification: 200×. The scale bar represents 100 μm. The total number of SMA+ blood vessels of selected bone biopsies taken from control sides without stem cells (white bars; $n = 3$), and study sides with stem cells (black bars; $n = 5$) from patients treated with β-TCP (**b**) or BCP (**e**). β-TCP, β-tricalcium phosphate; BCP, biphasic calcium phosphate; NB, native bone; TZ, transition zone; SVF, stromal vascular fraction; tot #, total number.

When analyzing the relative percentages of SMA+ small and large blood vessels in the different regions of the scaffold area (Table 2), we observed in the control sides of the β-TCP group equal levels of small and large vessels in regions I and II, and a sharp drop of both types of vessels in region III.

In the SVF-supplemented sides of this group, a similar but lower relative percentage was found, but now with a 2–3 fold increase in region III. In the control sides of the BCP group, blood vessels were only observed in regions I and III, in a 3–4:1 ratio. In contrast, the SVF-supplemented sides of the BCP group showed a divergent pattern, i.e., in regions I and II equal levels of small and large vessels were seen, but in region III a strong increase in large vessels and concomitant decrease in smaller vessels was observed, indicating more mature vessel formation in region III upon SVF supplementation.

When analyzing the total blood vessel numbers (Figure 2d,h) and the subdivision in small and large vessels (Figure 5a,b,g,h) of the control and SVF-supplemented sides per biomaterial in each region, no significant differences were observed.

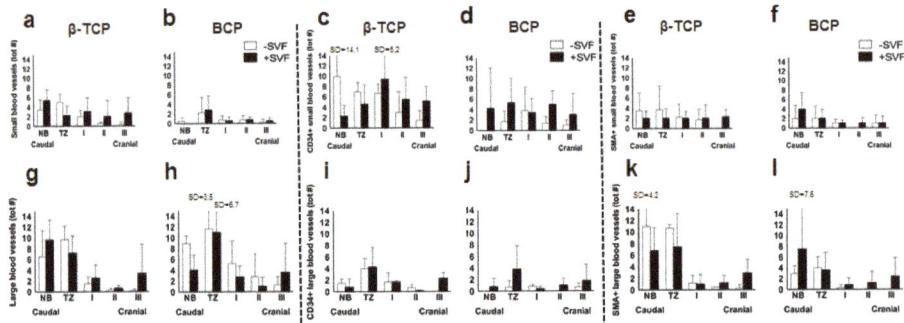

Figure 5. Histomorphometrical analysis of the absolute number of small and large blood vessels from maxillary bone biopsies in patients treated with β-TCP (a,g) and BCP (b,h) with stem cells (black bars, $n = 5$) and without stem cells (white bars, $n = 3$). The total number of CD34+ small and large blood vessels from maxillary bone biopsies in patients treated with β-TCP (c,i) and BCP (d,j) with stem cells (black bars, $n = 5$) and without stem cells (white bars, $n = 3$). The total number of SMA+ small and large blood vessels from maxillary bone biopsies in patients treated with β-TCP (e,k) and BCP (f,l) with stem cells (black bars, $n = 5$) and without stem cells (white bars, $n = 3$). See the legend to Figure 2 for the definitions of regions I-III. β-TCP, β-tricalcium phosphate; BCP, biphasic calcium phosphate; NB, Native bone; TZ, transition zone; SVF, stromal vascular fraction; tot #, total number.

A comparison of the CD34+ blood vessel total numbers (Figure 3c,f) and the subdivision in small and large vessels (Figure 5c,d,i,j) of the control and SVF-supplemented sides per biomaterial in each region showed no significant differences. When comparing the two biomaterials for their effect on CD34+ vasculogenesis in the presence or absence of SVF, it could be concluded that SVF supplementation significantly promoted CD34+ large vessel formation in region I in β-TCP, but not BCP ($p = 0.017$). In the absence of SVF, no difference between the two biomaterials was observed.

Comparison of the SMA+ blood vessel total numbers (Figure 4c,f) and the subdivision in small and large vessels (Figure 5e,f,k,l) of the control and SVF-supplemented sides per biomaterial in each region showed no significant differences. When analyzing the two biomaterials for their effect on SMA+ vasculogenesis in the presence or absence of SVF, it could be concluded that β-TCP in the absence of SVF significantly promoted total SMA+ vessel numbers ($p = 0.037$) and, in particular, large SMA+ vessels ($p = 0.025$) in region II when compared to BCP. In the presence of SVF, no difference between the two biomaterials was observed.

4. Discussion

In bone tissue engineering adequate vascularization is crucial for timely and adequate transport of nutrients and waste removal, and the provision of progenitor cells for tissue remodeling and repair. It is widely agreed that vascularization and bone formation are highly linked, and that vascularization precedes osteogenesis during both embryonic development and adult bone healing [38]. Since we hypothesized that supplementation of bone substitutes with SVF, containing adipose stem cells, will positively contribute not only to bone formation, as shown previously [27], but likely also to vascularization, the current study aimed to evaluate vascularization (number and size of blood vessels) in relation to bone formation potential of SVF.

Histomorphometric analysis and quantitative assessments of CD34- and SMA-immunostained blood vessels showed: (i) a clear trend towards increased bone formation in the SVF-supplemented group vs. the control group, in particular in the most cranial part (region III) of the biopsies, which was significantly higher in the β-TCP vs. the BCP group; (ii) comparable patterns of angiogenic and bone formation levels; (iii) comparable total blood vessel numbers obtained from histomorphometric and immunohistochemical analyses; (iv) SVF significantly promoted CD34+ large vessel formation in region I in β-TCP, but not BCP; (v) in the BCP group supplemented with SVF, region II displayed considerably higher CD34+ small and large vessel numbers; and (vi) in the absence of SVF the β-TCP, but not the BCP scaffolds, promoted both small and large vessel numbers in region II.

Our data analysis of the events occurring in the graft area using a division in regions provides important additional insight in course of events and local differences which would otherwise have been masked in a more global evaluation of the total graft area as a whole. The striking observation of the far more active bone formation observed at the cranial side of the biopsies in the case of SVF supplementation (7/10 biopsies) compared to the control side (1/6), as reported before [27], appears to match our current finding of higher blood vessel counts and the presence of more mature vessels (amongst others characterized by SMA+ staining) in these areas of active bone formation. The strong angiogenic potential of SVF has already been reported in numerous other studies [26,30,39]. The current findings in the six-month human model also adds to our earlier findings in previous preclinical large animal studies of our group in which we found that already after one month, SVF generates larger, more mature vessels compared to non-supplemented scaffolds [40]. The diameter of the blood vessels (mean diameter 30 µm) is well above the size of capillaries (5–10 µm) and can, therefore, be considered sufficient for bone formation [40]. Unfortunately, our study setup did not allow conclusions on blood vessel orientation, for it would have been interesting to evaluate whether the vessels would align in the principal loading direction.

Future studies with inclusion of a higher number of patients might reveal significant differences between groups where significance could not be reached due to the relatively low number of patients included in the current study. It would be interesting to also determine the orientation of the blood vessels based on an analysis of the blood vessel cross-sectional area, to elucidate if the orientation of blood vessels may be different in the scaffold area of stem cell-treated patients compared to controls.

5. Conclusions

SVF, and the stem cells residing therein, ASCs, have shown high angiogenic potential, making them highly interesting for tissue regeneration in the oral and maxillofacial area, but also for other clinical disciplines. The maxillary sinus floor elevation model is unique by allowing histological examination of biopsies removed prior to dental implant placement without interfering with the clinical routine, and intra-patient treatment comparisons when using a "split-mouth" design. Within these biopsies, we were able to demonstrate an increase in the number and maturity of blood vessels, in particular in the most cranial part, when patients were treated with stem cells. Bone percentages seem to correlate with blood vessel formation and are higher in study versus control biopsies in the cranial area, in particular in β-tricalcium phosphate-treated patients. To our knowledge, this is the first study directly linking SVF-induced bone formation and blood vessel formation in a clinical setting. The pro-angiogenic, bone formation-enhancing effects of SVF provide great potential for clinical bone tissue engineering.

Supplementary Materials: The following are available online at www.mdpi.com/1996-1944/11/1/161/s1, Table S1. Percentages of small and large blood vessels from maxillary bone biopsies obtained from patients treated with β-TCP and BCP without stem cells (−SVF, n = 3) and with stem cells (+SVF, n = 5) by histomorphometrical and immunohistochemical analysis.

Acknowledgments: The study was supported by the International Team for Implantology (ITI), grant number 1079_2015. We would like to thank Henk Jan Prins for his valuable scientific input, and Marion van Duin and Huib van Essen for their excellent technical support, and Irene Aartman for help with the statistical analysis.

Author Contributions: Elisabet Farré-Guasch (E.F.G.), Nathalie Bravenboer (N.B.), Marco N. Helder (M.N.H.), Engelbert A.J.M. Schulten (E.A.J.M.S.), Christiaan M. ten Bruggenkate (C.M.t.K.), and Jenneke Klein-Nulend (J.K.N.) conceived and designed the experiments; E.F.G. performed the experiments; all authors analyzed the data; E.F.G., M.N.H., E.A.J.M.S., C.M.t.K., and J.K.N. drafted the manuscript; and all authors performed the critical revision of the article and approved the final, submitted version.

Conflicts of Interest: The authors declare no conflict of interest.

References

1. Tatum, H., Jr. Maxillary and sinus implant reconstructions. *Dent. Clin. N. Am.* **1986**, *30*, 207–229. [CrossRef] [PubMed]
2. Zijderveld, S.A.; ten Bruggenkate, C.M.; van Den Bergh, J.P.; Schulten, E.A.J.M. Fractures of the iliac crest after split-thickness bone grafting for preprosthetic surgery: Report of 3 cases and review of the literature. *J. Oral Maxillofac. Surg.* **2004**, *62*, 781–786. [CrossRef] [PubMed]
3. Vamze, J.; Pilmane, M.; Skagers, A. Biocompatibility of pure and mixed hydroxyapatite and α-tricalcium phosphate implanted in rabbit bone. *J. Mater. Sci. Mater. Med.* **2015**, *26*, 73. [CrossRef] [PubMed]
4. Al-Sanabani, J.S.; Madfa, A.A.; Al-Sanabani, F.A. Application of calcium phosphate materials in dentistry. *Int. J. Biomater.* **2013**, *2013*, 1–12. [CrossRef] [PubMed]
5. Sheikh, Z.; Abdallah, M.N.; Hanafi, A.A.; Misbahuddin, S.; Rashid, H.; Glogauer, M. Mechanisms of in vivo degradation and resorption of calcium phosphate based biomaterials. *Materials* **2015**, *8*, 7913–7925. [CrossRef] [PubMed]
6. Zerbo, I.R.; Zijderveld, S.A.; de Boer, A.; Bronckers, A.L.J.J.; de Lange, G.; ten Bruggenkate, C.M.; Burger, E.H. Histomorphometry of human sinus floor augmentation using a porous beta-tricalcium phosphate: A prospective study. *Clin. Oral Implants Res.* **2004**, *15*, 724–732. [CrossRef] [PubMed]
7. Zijderveld, S.A.; Giltaij, L.R.; Bergh, J.P.; Smit, T.H. Pre-clinical and clinical experiences with BMP-2 and BMP-7 in sinus floor elevation surgery. A comparison. *J. Musculoskelet. Res.* **2002**, *6*, 43–54. [CrossRef]
8. Zijderveld, S.A.; Zerbo, I.R.; van den Bergh, J.P.; Schulten, E.A.J.M.; ten Bruggenkate, C.M. Maxillary sinus floor augmentation using a beta-tricalcium phosphate (Cerasorb) alone compared to autogenous bone grafts. *Int. J. Oral Maxillofac. Implants* **2005**, *20*, 432–440. [PubMed]
9. Li, Y.; Jiang, T.; Zheng, L.; Zhao, J. Osteogenic differentiation of mesenchymal stem cells (MSCs) induced by three calcium phosphate ceramic (CaP) powders: A comparative study. *Mater. Sci. Eng. C Mater. Biol. Appl.* **2017**, *80*, 296–300. [CrossRef] [PubMed]
10. Urquia Edreira, E.R.; Hayrapetyan, A.; Wolke, J.G.; Croes, H.J.; Klymov, A.; Jansen, J.A.; van den Beucken, J.J. Effect of calcium phosphate ceramic substrate geometry on mesenchymal stromal cell organization and osteogenic differentiation. *Biofabrication* **2016**, *8*. [CrossRef] [PubMed]
11. Ribatti, D.; Crivellato, E. "Sprouting angiogenesis", a reappraisal. *Dev. Biol.* **2012**, *372*, 157–165. [CrossRef] [PubMed]
12. Logsdon, E.A.; Finley, S.D.; Popel, A.S.; Mac Gabhann, F. A systems biology view of blood vessel growth and remodelling. *J. Cell. Mol. Med.* **2014**, *18*, 1491–1508. [CrossRef] [PubMed]
13. Bergers, G.; Song, S. The role of pericytes in blood-vessel formation and maintenance. *Neuro Oncol.* **2005**, *7*, 452–464. [CrossRef] [PubMed]
14. Colton, C.K. Implantable biohybrid artificial organs. *Cell Transplant.* **1995**, *4*, 415–436. [CrossRef] [PubMed]
15. Moioli, E.K.; Clark, P.A.; Chen, M.; Dennis, J.E.; Erickson, H.P.; Gerson, S.L.; Mao, J.J. Synergistic actions of hematopoietic and mesenchymal stem/ progenitor cells in vascularizing bioengineered tissues. *PLoS ONE* **2008**, *3*, e3922. [CrossRef] [PubMed]
16. Farré-Guasch, E.; Martí-Pagès, C.; Hernández-Alfaro, F.; Klein-Nulend, J.; Casals, N. Buccal fat pad, an oral access source of human adipose stem cells with potential for osteochondral tissue engineering: An in vitro study. *Tissue Eng. Part C Methods* **2010**, *16*, 1083–1094. [CrossRef] [PubMed]

17. Aust, L.; Devlin, B.; Foster, S.J.; Halvorsen, Y.D.; Hicok, K.; du Laney, T.; Sen, A.; Willingmyre, G.D.; Gimble, J.M. Yield of human adipose-derived adult stem cells from liposuction aspirates. *Cytotherapy* **2004**, *6*, 7–14. [CrossRef] [PubMed]
18. Helder, M.N.; Knippenberg, M.; Klein-Nulend, J.; Wuisman, P.I.J.M. Stem cells from adipose tissue allow challenging new concepts for regenerative medicine. *Tissue Eng.* **2007**, *13*, 1799–1808. [CrossRef] [PubMed]
19. Zuk, P.A.; Zhu, M.; Mizuno, H.; Huang, J.; Futrell, J.W.; Katz, A.J.; Benhaim, P.; Lorenz, H.P.; Hedrick, M.H. Multilineage cells from human adipose tissue: Implications for cell-based therapies. *Tissue Eng.* **2001**, *7*, 211–228. [CrossRef] [PubMed]
20. Agata, H.; Asahina, I.; Watanabe, N.; Ishii, Y.; Kubo, N.; Ohshima, S.; Yamazaki, M.; Tojo, A.; Kagami, H. Characteristic change and loss of in vivo osteogenic abilities of human bone marrow stromal cells during passage. *Tissue Eng. Part A* **2010**, *16*, 663–673. [CrossRef] [PubMed]
21. Izadpanah, R.; Kaushal, D.; Kriedt, C.; Tsien, F.; Patel, B.; Dufour, J.; Bunnell, B.A. Long-term in vitro expansion alters the biology of adult mesenchymal stem cells. *Cancer Res.* **2008**, *68*, 4229–4238. [CrossRef] [PubMed]
22. Rubio, D.; Garcia, S.; Paz, M.F.; De la Cueva, T.; Lopez-Fernandez, L.A.; Lloyd, A.C.; Garcia-Castro, J.; Bernad, A. Molecular characterization of spontaneous mesenchymal stem cell transformation. *PLoS ONE* **2008**, *3*, e1398. [CrossRef] [PubMed]
23. Cho, C.H.; Koh, Y.J.; Han, J.; Sung, H.K.; Jong Lee, H.; Morisada, T.; Schwendener, R.A.; Brekken, R.A.; Kang, G.; Oike, Y.; et al. Angiogenic role of LYVE-1-positive macrophages in adipose tissue. *Circ. Res.* **2007**, *100*, e47–e57. [CrossRef] [PubMed]
24. Overman, J.R.; Farré-Guasch, E.; Helder, M.N.; ten Bruggenkate, C.M.; Schulten, E.A.J.M.; Klein-Nulend, J. Short (15 min) bone morphogenetic protein-2 treatment stimulates osteogenic differentiation of human adipose stem cells seeded on calcium phosphate scaffolds in vitro. *Tissue Eng. Part A* **2013**, *19*, 571–581. [CrossRef] [PubMed]
25. Overman, J.R.; Helder, M.N.; ten Bruggenkate, C.M.; Schulten, E.A.J.M.; Klein-Nulend, J.; Bakker, A.D. Growth factor gene expression profiles of bone morphogenetic protein-2-treated human adipose stem cells seeded on calcium phosphate scaffolds in vitro. *Biochimie* **2013**, *95*, 2304–2313. [CrossRef] [PubMed]
26. Kim, A.; Kim, D.H.; Song, H.R.; Kang, W.H.; Kim, H.J.; Lim, H.C.; Cho, D.W.; Bae, J.H. Repair of rabbit ulna segmental bone defect using freshly isolated adipose-derived stromal vascular fraction. *Cytotherapy* **2012**, *14*, 296–305. [CrossRef] [PubMed]
27. Prins, H.J.; Schulten, E.A.J.M.; Ten Bruggenkate, C.M.; Klein-Nulend, J.; Helder, M.N. Bone regeneration using the freshly isolated autologous stromal vascular fraction of adipose tissue in combination with calcium phosphate ceramics. *Stem Cells Transl. Med.* **2016**, *5*, 1362–1374. [CrossRef] [PubMed]
28. Ko, Y.J.; Koh, B.I.; Kim, H.; Joo, H.J.; Jin, H.K.; Jeon, J.; Choi, C.; Lee, D.H.; Chung, J.H.; Cho, C.H.; et al. Stromal vascular fraction from adipose tissue forms profound vascular network through the dynamic reassembly of blood endothelial cells. *Arterioscler. Thromb. Vasc. Biol.* **2011**, *31*, 1141–1150. [CrossRef]
29. Rubina, K.; Kalinina, N.; Efimenko, A.; Rubina, K.; Kalinina, N.; Efimenko, A.; Lopatina, T.; Melikhova, V.; Tsokolaeva, Z.; Sysoeva, V.; et al. Adipose stromal cells stimulate angiogenesis via promoting progenitor cell differentiation, secretion of angiogenic factors, and enhancing vessel maturation. *Tissue Eng. Part A* **2009**, *15*, 2039–2050. [CrossRef] [PubMed]
30. Madonna, R.; De Caterina, R. In vitro neovasculogenic potential of resident adipose tissue precursors. *Am. J. Physiol. Cell Physiol.* **2008**, *295*, 1271–1280. [CrossRef] [PubMed]
31. Oostlander, A.E.; Bravenboer, N.; Sohl, E.; Holzmann, P.J.; van der Woude, C.J.; Dijkstra, G.; Stokkers, P.C.; Oldenburg, B.; Netelenbos, J.C.; Hommes, D.W.; et al. Histomorphometric analysis reveals reduced bone mass and bone formation in patients with quiescent Crohn's disease. Dutch Initiative on Crohn and Colitis (ICC). *Gastroenterology* **2011**, *140*, 116–123. [CrossRef] [PubMed]
32. Schulten, E.A.J.M.; Prins, H.J.; Overman, J.R.; Helder, M.N.; ten Bruggenkate, C.M.; Klein-Nulend, J. A novel approach revealing the effect of a collagenous membrane on osteoconduction in maxillary sinus floor elevation with beta-tricalcium phosphate. *Eur. Cells Mater.* **2013**, *25*, 215–228. [CrossRef]
33. Parfitt, A.M.; Drezner, M.K.; Glorieux, F.H.; Kanis, J.A.; Malluche, H.; Meunier, P.J.; Ott, S.M.; Recker, R.R. Bone histomorphometry: Standardization of nomenclature, symbols, and units. Report of the ASBMR Histomorphometry Nomenclature Committee. *J. Bone Miner. Res.* **1987**, *2*, 595–610. [CrossRef] [PubMed]

34. Chen, Y.; Wang, J.; Zhu, X.D.; Tang, Z.R.; Yang, X.; Tan, Y.F.; Fan, Y.J.; Zhang, X.D. Enhanced effect of β-tricalcium phosphate phase on neovascularization of porous calcium phosphate ceramics: In vitro and in vivo evidence. *Acta Biomater.* **2015**, *11*, 435–448. [CrossRef] [PubMed]
35. Pusztaszeri, M.P.; Seelentag, W.; Bosman, F.T. Immunohistochemical expression of endothelial markers CD31, CD34, von Willebrand Factor, and Fli-1 in normal human tissues. *J. Histochem. Cytochem.* **2006**, *54*, 385–395. [CrossRef] [PubMed]
36. Brey, E.M.; McIntire, L.V.; Johnston, C.M.; Reece, G.P.; Patrick, C.W., Jr. Three-dimensional, quantitative analysis of desmin and smooth muscle alpha actin expression during angiogenesis. *Ann. Biomed. Eng.* **2004**, *32*, 1100–1107. [CrossRef] [PubMed]
37. Amir, L.R.; Becking, A.G.; Jovanovic, A.; Perdijk, F.B.; Everts, V.; Bronckers, A.L.J.J. Formation of new bone during vertical distraction osteogenesis of the human mandible is related to the presence of blood vessels. *Clin. Oral Implants Res.* **2006**, *17*, 410–416. [CrossRef] [PubMed]
38. Roux, B.M.; Cheng, M.; Brey, E.M. Engineering clinically relevant volumes of vascularized bone. *J. Cell. Mol. Med.* **2015**, *19*, 903–914. [CrossRef] [PubMed]
39. Planat-Benard, V.; Silvestre, J.S.; Cousin, B.; André, M.; Nibbelink, M.; Tamarat, R.; Clergue, M.; Manneville, C.; Saillan-Barreau, C.; Duriez, M.; et al. Plasticity of human adipose lineage cells toward endothelial cells: Physiological and therapeutic perspectives. *Circulation* **2004**, *109*, 656–663. [CrossRef] [PubMed]
40. Vergroesen, P.P.; Kroeze, R.J.; Helder, M.N.; Smit, T.H. The use poly(L-lactide-co-caprolactone) as a scaffold for adipose stem cells in bone tissue engineering: Application in a spinal fusion model. *Macromol. Biosci.* **2011**, *14*, 722–730. [CrossRef] [PubMed]

© 2018 by the authors. Licensee MDPI, Basel, Switzerland. This article is an open access article distributed under the terms and conditions of the Creative Commons Attribution (CC BY) license (http://creativecommons.org/licenses/by/4.0/).

Article

Enhancement of Osteoblastic-Like Cell Activity by Glow Discharge Plasma Surface Modified Hydroxyapatite/β-Tricalcium Phosphate Bone Substitute

Eisner Salamanca [1], Yu-Hwa Pan [1,2,3,4], Aileen I. Tsai [2], Pei-Ying Lin [1], Ching-Kai Lin [2], Haw-Ming Huang [1,5], Nai-Chia Teng [1,6], Peter D. Wang [1,6,*] and Wei-Jen Chang [1,7,*]

1 School of Dentistry, College of Oral Medicine, Taipei Medical University, Taipei 110, Taiwan; eisnergab@hotmail.com (E.S.); shalom.dc@msa.hinet.net (Y.-H.P.); payinglin53@gmail.com (P.-Y.L.); hhm@tmu.edu.tw (H.-M.H.); dianaten@tmu.edu.tw (N.-C.T.)
2 Department of Dentistry, Chang Gung Memorial Hospital, Taipei 105, Taiwan; ait001@adm.cgmh.org.tw (A.I.T.); philipcklin@msn.com (C.-K.L.)
3 Graduate Institute of Dental & Craniofacial Science, Chang Gung University, Taoyuan 333, Taiwan
4 School of Dentistry, College of Medicine, China Medical University, Taichung 404, Taiwan
5 Graduate Institute of Biomedical Materials & Tissue Engineering, College of Oral Medicine, Taipei Medical University, Taipei 110, Taiwan
6 Dental Department, Taipei Medical University Hospital, Taipei 110, Taiwan
7 Dental Department, Taipei Medical University, Shuang-Ho Hospital, Taipei 235, Taiwan
* Correspondence: dpw1@tmu.edu.tw (P.D.W.); cweijen1@tmu.edu.tw (W.-J.C.);
  Tel.: +886-2-2736-1661 (ext. 5148) (P.D.W. & W.-J.C.); Fax: +886-2-2736-2295 (P.D.W. & W.-J.C.)

Received: 5 October 2017; Accepted: 21 November 2017; Published: 23 November 2017

Abstract: Glow discharge plasma (GDP) treatments of biomaterials, such as hydroxyapatite/β-tricalcium phosphate (HA/β-TCP) composites, produce surfaces with fewer contaminants and may facilitate cell attachment and enhance bone regeneration. Thus, in this study we used argon glow discharge plasma (Ar-GDP) treatments to modify HA/β-TCP particle surfaces and investigated the physical and chemical properties of the resulting particles (HA/β-TCP + Ar-GDP). The HA/β-TCP particles were treated with GDP for 15 min in argon gas at room temperature under the following conditions: power: 80 W; frequency: 13.56 MHz; pressure: 100 mTorr. Scanning electron microscope (SEM) observations showed similar rough surfaces of HA/β-TCP + Ar-GDP HA/β-TCP particles, and energy dispersive spectrometry analyses showed that HA/β-TCP surfaces had more contaminants than HA/β-TCP + Ar-GDP surfaces. Ca/P mole ratios in HA/β-TCP and HA/β-TCP + Ar-GDP were 1.34 and 1.58, respectively. Both biomaterials presented maximal intensities of X-ray diffraction patterns at 27° with 600 a.u. At 25° and 40°, HA/β-TCP + Ar-GDP and HA/β-TCP particles had peaks of 200 a.u., which are similar to XRD intensities of human bone. In subsequent comparisons, MG-63 cell viability and differentiation into osteoblast-like cells were assessed on HA/β-TCP and HA/β-TCP + Ar-GDP surfaces, and Ar-GDP treatments led to improved cell growth and alkaline phosphatase activities. The present data indicate that GDP surface treatment modified HA/β-TCP surfaces by eliminating contaminants, and the resulting graft material enhanced bone regeneration.

Keywords: HA/β-TCP; argon glow discharge plasma; guided bone regeneration; osteoconduction; cell viability; differentiation

1. Introduction

Multiple regenerative procedures have been developed for the treatment of deep infrabony defects, furcation involvements, and for socket preservation after tooth extraction [1–4]. Autogenous bone grafts remain the gold standard for bone regeneration procedures because they contain viable osteoblasts, organic and inorganic matrices, and biological modifiers. Although autogenous bone grafts are ideal for hard-tissue grafts, they are disadvantaged by limited availability, the tendency toward partial resorption, the requirement for additional surgery, and the ensuing increases in morbidity [5]. Thus, further studies are urgently required to develop and compare alternatives to allogenic grafts, including optional biomaterials such as xenogenic grafts from the same and other species, alloplastic materials, and synthetic and inorganic implant materials [2,6]. These materials may provide scaffolds for bone formation (osteoconduction) and could contain bone-forming cells (osteogenesis) or bone-inductive substances (osteoinduction) [3]. However, it remains unclear which graft materials are the most suitable for bone regeneration [7–9].

In 2005, Trombelli et al. showed that various forms of hydroxyapatite (HA) significantly improve clinical attachment levels compared with periodontal surgery [2]. Moreover, in 2003, Reynolds et al. reported good clinical outcomes following the use of calcium phosphate ceramic in periodontal therapy [6]. Subsequently, biphasic calcium phosphates (BCPs) comprising mixtures of HA and beta tricalcium phosphate (β-TCP) at varying ratios were well documented as bioinert and bioactive alloplastic materials [10,11]. In particular, the ratio of approximately 60% HA to 40% β-TCP produced optimal resorption by the material and maintenance of osteoconductive properties [12–15].

Glow discharge plasma (GDP) is formed by the passage of electric current through a low-pressure gas and is widely used for cleaning, sterilizing, and modifying biomaterial surfaces [16–21]. Glow-discharge techniques are well established cleaning methods in the microelectronics industry and are under current consideration in the production of biomaterials [22,23]. Glow-discharge methods offer great advantages with respect to the possible range of modifications. Using appropriate GDP parameters with argon, plasmas can remove all traces of potentially problematic entities from biomaterial surfaces, including contaminants, impurities, and native oxide layers [16], warranting further characterization of sputtered HA/β-TCP biomaterial surfaces after GDP treatment.

Bone scaffold can be manufactured by several synthetic routes developed to prepare BCP bioceramics of variable HA/β-TCP ratios simulating the physical and biological properties of natural bones [24]. Multiple techniques, such as gas foaming, freeze drying, thermally induced phase separation, precipitation, hydrolysis, mechanical mixture, among others [25]. While using these manufacturing techniques few papers have reported biomaterials sterilization processes, storage conditions, and the impurities that cover the materials [24]. Indeed, HA/β-TCP surface impurities removal using Argon plasma sputtering, has not been well documented and needs to be better understood.

Herein, a surface modification of HA/β-TCP particles using GDP treatments and an evaluation of enhancements in bone regeneration properties were carried out in the present study. Specifically, we determined the physical and chemical properties of HA/β-TCP particles after GDP surface treatment (HA/β-TCP + Ar-GDP) and tested the resulting biological effects on MG-63 cell viability and differentiation into osteoblast like cells.

2. Materials and Methods

2.1. Sample Preparation

Bone substitute granules of 500–1000 µm biphasic ceramic materials comprising 60% HA and 40% β-TCP (MBCP) were purchased from Biomatlante (Vigneux-de-Bretagne, France).

2.2. GDP Sputtering

HA/β-TCP particles were treated for 15 min with GDP (PJ; AST Products Inc., North Billerica, MA, USA) in the presence of argon gas at room temperature under the following conditions: power: 80 W; frequency: 13.56 MHz; pressure: 100 mTorr (Figure 1).

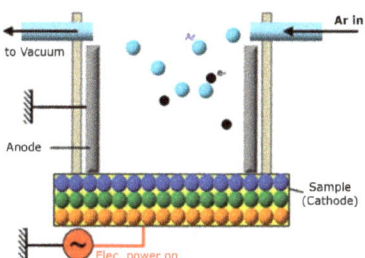

Figure 1. Glow discharge plasma design.

2.3. Surface Topography Evaluations

Surface morphologies of HA/β-TCP + Ar-GDP were observed and compared with those of HA/β-TCP using 30 scanning electron microscopy images (SEM; Model 2400; Hitachi, Ltd., Tokyo, Japan) from the same sample. Prior to imaging, 25 nm thick layers of palladium gold were sputter-coated onto a sample using a sputtering apparatus (IB-2; Hitachi, Ltd., Tokyo, Japan). The evaluation was conducted with one sample only ($n = 1$).

2.4. Energy Dispersive Spectrometry

Elemental analyses of HA/β-TCP + Ar-GDP samples were performed using the same SEM apparatus described above coupled with an energy dispersive X-ray spectrometer (EDS; Model 2400; Hitachi, Ltd., Tokyo, Japan) ($n = 1$).

2.5. X-ray Photoelectron Spectroscopy (XPS)

Elemental and chemical analyses of treated surfaces were performed using X-ray photoelectron spectroscopy (ESCA system; VG Scientific, West Sussex, UK) with a 1486.6 eV monochromatic Al X-ray source. Spectra were collected at a normalized electron take-off angle to the sample surfaces. High resolution [C1s], [O1s], [Ca2p], and [P2p] spectra were obtained from 25 surface-treated samples ($n = 1$).

2.6. X-ray Diffraction Analyses

Crystalline structures and chemical compositions using approximately 0.5 mg of HA/β-TCP + Ar-GDP and HA/β-TCP, were analyzed using powder X-ray diffraction (XRD, X'Pert³ Powder, PANalytical Co. Ltd., Almelo, The Netherlands), at 40 kV and 40 mA with a scanning speed of 0.5°/s and a scanning range of 20°–50° ($n = 1$).

2.7. Cell Viability Assays

Cytotoxicity was assessed in vitro according to international standards for the evaluation of dental materials (ISO 7405:2008) Cytotoxicity was assessed in vitro according to international standards for the evaluation of dental materials (ISO 7405:2008) Cytotoxicity was assessed in vitro according to international standards for the evaluation of dental materials (ISO 7405:2008) Cytotoxicity was assessed in vitro according to international standards for the evaluation of dental materials (ISO 10993-5:2009 [26]). Passage 17 of the MG-63 osteoblast-like cells was seeded into 24-well Petri dishes (Nunclon; Nunc, Roskilde, Denmark) at a density of 1×10^4 cells/mL using 1 mL of Dulbecco's modified Eagle's medium (DMEM; HyClone, Logan, UT, USA) supplemented with L-glutamine (4 mmol/L), 10% fetal bovine serum (FBS), and 1% penicillin streptomycin in every well for 24 h. The media was then removed and substituted with a new media consisting of the previous described DMEM + HA/β-TCP + Ar-GDP or HA/β-TCP particles in a concentration of 1 g/10 mL using just the media for the test wells. Control wells used the same DMEM media first described without any particle graft. Cells were cultured over periods of 1, 3, and 5 days at 5% CO_2, 37 °C and 100% humidity. The 0 h starting point was defined as the moment when DMEM + particle grafts were added to the test wells.

At Days 1, 3, and 5, cell viabilities were assessed according to metabolic activities using (3-4,5-dimethylethiazol-2-yl)-2,5-diphenyl tetrazolium bromide (MTT) reduction assays. In these experiments, MTT was metabolically reduced by mitochondrial dehydrogenase in viable cells and the resulting production of colored formazan was determined according to the manufacturer's instructions (MTT kit, Roche Applied Science, Mannheim, Germany). After removing MTT containing media from cells, formazan crystals were dissolved in 500 μL of dimethyl sulfoxide for 5 min, and optical density absorbance was recorded using an enzyme linked immunosorbent assay (ELISA) reader at 570 nm. Results were expressed as percentages ($n = 6$).

2.8. Cell Morphology

To investigate cell morphology on biomaterial particles, cells were cultured into 24-well Petri dishes (Nunclon; Nunc, Roskilde, Denmark) at a density of 1×10^4 cells/mL osteogenic inductive media, which was a combination of the media described in the cell viability assay plus HA/β-TCP + Ar-GDP or HA/β-TCP particles in a concentration of 1 g/10 mL using 1 mL of media and particle grafts. Images were recorded at 40× magnification on Days 1, 3, and 5 using an optical microscope (Olympus BH-2, Tokyo, Japan) with a camera (SPOT™ idea, SPOT Imaging, Sterling Heights, MI, USA) and were analyzed using the corresponding software (SPOT imaging software, Sterling Heights, MI, USA) ($n = 6$).

2.9. Alkaline Phosphatase Assays

MG-63 cells of Passage 18 were cultured following the cell viability assay protocol for Days 1, 3, and 5. Once they reach the timepoints, cells in the wells were washed twice with phosphate-buffered saline (PBS). PBS was suctioned and 300 μL of Triton-100 at 0.05% was added. Cells went through 3 cycles, each one of 5 min at 37 °C and −4 °C for cell rupture. Later, samples were placed into 96 wells. ALP activities were determined using p-nitrophenyl phosphate (pNPP; Sigma, St. Louis, MO, USA) as the substrate. Enzyme activities were quantified from absorbance measurements at 405 nm using Multiskan™ GO Microplate Spectrophotometer (Thermo Fisher Scientific, Waltham, MA, USA) and were normalized to total protein contents, which were determined using the BCA method in aliquots of the same samples with a Pierce (Rockford, IL, USA) protein assay kit ($n = 6$).

2.10. Statistical Analyses

All values were expressed as mean ± standard deviation. Jarque-Bera was used to test the normality in the results. Differences between HA/β-TCP + Ar-GDP and HA/β-TCP groups were identified using Student's *t*-tests and were considered significant when $p < 0.05$.

3. Results

3.1. SEM Observations

Morphological structures of HA/β-TCP + Ar-GDP and HA/β-TCP surfaces were similar. SEM analyses revealed no discernible damage associated with the surface treatment, and HA/β-TCP + Ar-GDP resembled HA/β-TCP, with rough surfaces within macro- and microporous structures that induce osteoblastic cell attachment and allow for the passage of fluids (Figure 2).

Structural characteristic groups and vibration bonds were examined by Fourier Transform Infrared Spectrum (FTIR) and are presented in the supplementary materials.

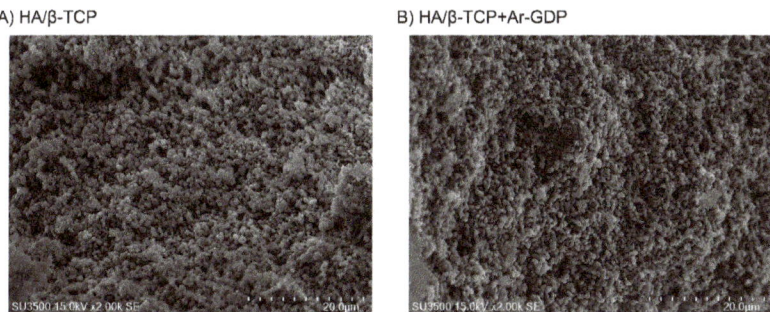

Figure 2. SEM images. (**A**) HA/β-TCP composite (×2000); (**B**) HA/β-TCP composite with argon glow discharge plasma treatment (×2000).

3.2. EDS Analysis

EDS analyses showed greater numbers of elements in HA/β-TCP than in HA/β-TCP + Ar-GDP. Specifically, magnesium, potassium, and niobium, which sometimes exist as residual impurities on HA/β-TCP surfaces, were not present after GDP treatment on HA/β-TCP + Ar-GDP. Moreover, the Ca/P mole ratio of HA/β-TCP particles was 1.34, which is not close to the stoichiometric value of HAP (1.67), while that of HA/β-TCP + Ar-GDP was 1.58, which is closer to the stoichiometric value for HA (Table 1, Figure 3).

Table 1. EDS analyses of particles with and without argon glow discharge plasma (Ar-GDP) treatment.

Materials	HA/β-TCP		HA/β-TCP + Ar-GDP	
Element	Weight %	Atomic %	Weight %	Atomic %
O	33.81	58.15	25.02	48.13
P	18.19	16.16	19.19	19.07
Au	13.34	1.86	16.43	2.57
Ca	31.64	21.72	39.36	30.23
Mg	0.66	0.75	0	0
Nb	1.62	0.48	0	0
Na	0.73	0.87	0	0
Total	100	100	100	100

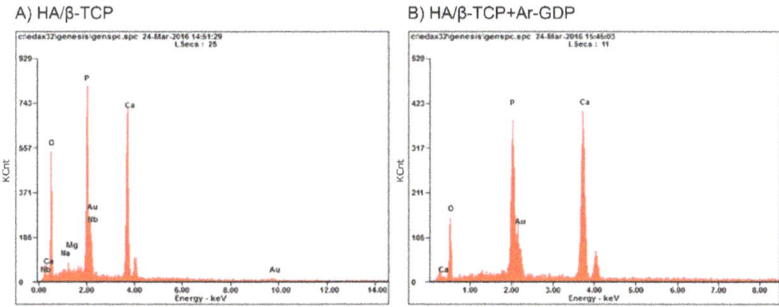

Figure 3. EDS spectra. The calcium and phosphate were observed on the surfaces of HA/β-TCP particles before (**A**) and after the glow discharge plasma treatment (**B**). However, the trace element impurities (Nb, Mg, and Na) were not observed after the glow discharge plasma treatment.

3.3. XPS

XPS determinations of the atomic surface concentrations of both materials are presented in Table 2, where similar elemental composition can be seen for both materials. However, the surface atomic composition percentage showed differences after GDP treatment, meaning surface modification had occurred after the argon glow discharge treatment. (Figure 4).

Table 2. XPS analyses of atomic compositions (%).

Materials	C1s	O1s	Na1s	P2p	Ca2p
HA/β-TCP	10.72	60.5	0.46	12.53	15.79
HA/β-TCP + Ar-GDP	12.06	59.79	0.2	12.06	16.09

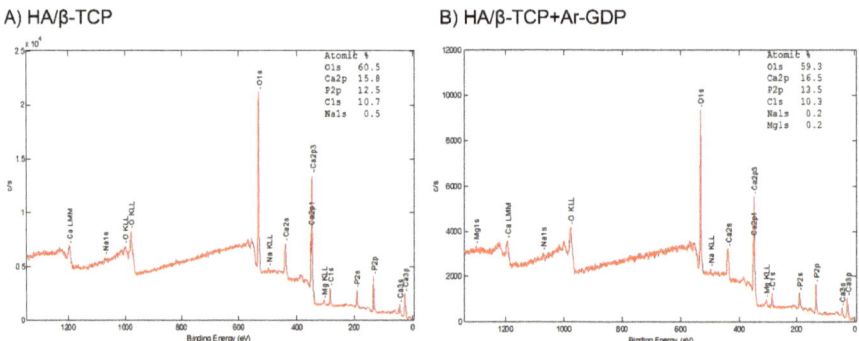

Figure 4. X-ray photoelectron spectroscopic analyses of atomic compositions (%). (**A**) HA/β-TCP composite (×2000); (**B**) HA/β-TCP composite with argon glow discharge plasma treatment. XPS results indicated that surface chemical bonding structure of calcium and phosphate had not changed after glow discharge plasma treatment.

XRD patterns showed no change in the pattern as a result of the Ar-GDP. Treated and non-treated bone substitutes had the highest intensity (600 a.u.) at 27°, and this value is similar to that for HA at 21°. At 25° and 40°, HA/β-TCP + Ar-GDP and HA/β-TCP had peaks of 200 a.u. (Figure 5).

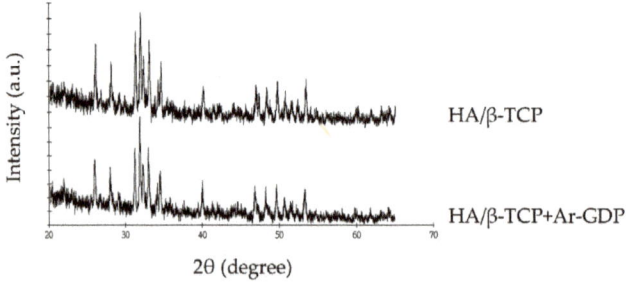

Figure 5. XRD patterns.

3.4. Cell Viability and Morphology Assessments

In MTT assays, both materials promoted osteoblastic cell proliferation, and confluent monolayers were present in all test wells, with similar pH values to those in controls. Cell viability on HA/β-TCP + Ar-GDP was 91.83% ± 1.58%, 93.46% ± 2.3%, and 106.43% ± 9.1% on Days 1, 3, and 5, respectively. In contrast, cell viability on HA/β-TCP surfaces were 93.58% ± 2.73%, 97.96% ± 3.82%, and 93.82% ± 7.1% on Days 1, 3, and 5, respectively (statistically significant difference $p < 0.05$, presented in Figure 6). MG-63 cells progressed from attachment to spreading on HA/β-TCP + Ar-GDP surfaces, leading to enhanced cell viability in comparison with that on HA/β-TCP surfaces on Day 5 (Figure 6).

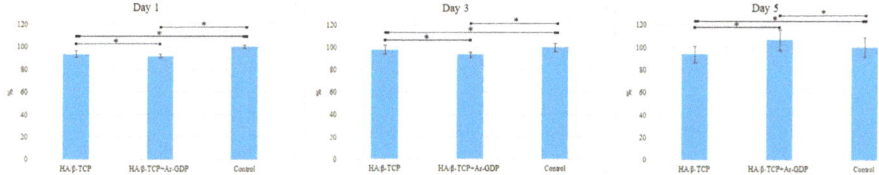

Figure 6. MTT assays on Days 1, 3, and 5. * Statistically significant difference ($p < 0.05$).

3.5. Cell Morphology

Morphology analyses of cells showed growth and spreading on both surfaces, with elongated appearances and the formation of relatively thin continuous monolayers. In the presence of biomaterials, cells surrounding the graft particles were observed on Days 1, 3, and 5 (Figure 7).

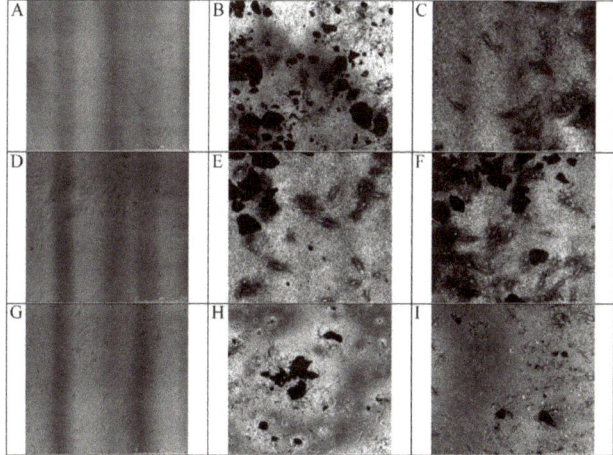

Figure 7. MG-63 cells co-cultured with particle grafts. Changes in MG-63 cell morphology on Days 1, 3, and 5. Magnification 40×. (**A**) Cells growing in Dulbecco's modified Eagle's medium on Day 1. (**B**) Cells growing with HA/β-TCP + Ar-GDP on Day 1. (**C**) Cells were cultured with HA/β-TCP and were grown for 1 day. (**D**) Cells in Dulbecco's modified Eagle's medium on Day 3. (**E**) Cells growing around HA/β-TCP + Ar-GDP on Day 3. (**F**) Cells growing around HA/β-TCP on Day 3. (**G**) Cells in Dulbecco's modified Eagle's medium on Day 5 showing a stage of development. (**H**) Cells with HA/β-TCP + Ar-GDP showing intimate contact with particle grafts on Day 5. (**I**) Cells on HA/β-TCP showed development and intimate contact with particle grafts on Day 5.

3.6. Alkaline Phosphatase Assay

Alkaline phosphatase tests showed that HA/β-TCP + Ar-GDP increased ALP activity in a time-dependent manner, with significantly greater enzyme activity compared to HA/β-TCP on Days 1, 3, and 5 (statistically significant difference $p < 0.05$, presented in Figure 8).

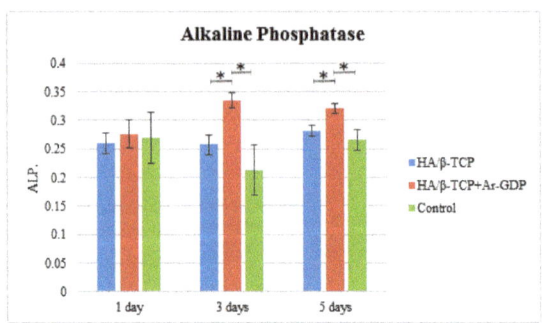

Figure 8. ALP assays. * Statistically significant difference ($p < 0.05$).

4. Discussion

The use of GDP as a technology for enhanced bone fusion and improved healing after skeletal injury has been investigated previously. Moreover, plasma treatments were previously used to hydrophilize surfaces of composite scaffolds and dental implants and facilitate cell adhesion [27–29]. Numerous studies show that the surface characteristics of biomaterials can directly influence cellular responses, thus affecting rates of growth and qualities of new tissue [30,31]. In the present study, we used Ar-GDP surface sputtering to remove element impurities from HA/β-TCP surfaces and enhance their bone regeneration properties. SEM analyses showed similar topographies of HA/β-TCP and HA/β-TCP + Ar-GDP surfaces, with rough surfaces within macro- and microporous structures that are ideal for bone regeneration and angiogenesis. The formation of a new vascular network, as Hing et al. explained, is strongly influenced by the degree of structural interconnectivity between pores, indicating that both micro- and macroporosity play a role in bone regeneration [32].

In addition, EDS analyses in the present study indicated HA/β-TCP Ca/P mole ratios lower than those previously reported of bone apatite crystals, which is due to the use of biphasic powders, where Ca/P mole ratios lower than 1.50 are easily obtained [33]. Apatite is considered calcium-deficient when the Ca/P mole ratios are lower than the stoichiometric value of 1.67 for pure calcium hydroxyapatite [34]. Though the Ca/P mole ratio of HA/β-TCP and HA/β-TCP + Ar-GDP in the present study was lower than the median reference value reported by Tzaphlidou et al. [35] in human rib bones (2.19), HA/β-TCP and HA/β-TCP + Ar-GDP allowed for cell attachment, proliferation, and expression. EDS analyses showed that HA/β-TCP + Ar-GDP had reduced contaminant concentrations. The use of XPS for surface characterization has shown this to be a technique that can provide all the information needed to distinguish the bulk composition from that of the outermost layers in particle grafts [36]. The influence of surface impurities on bioceramics is an important topic. As Franca et al. indicated in 2014, it is important to eliminate such impurities from bioceramics because they introduce critical defects, affecting their mechanical properties during lengthy implantation for bone regeneration [37].

In concordance with EDS and XPS analysis, MTT, morphology, and ALP assays, after 5 days, demonstrated superior properties for HA/β-TCP + Ar-GDP in contrast to those of HA/β-TCP, with improved cell proliferation, increased ALP activity, and enhanced differentiation into osteoblast like cells. One can attribute these differences to the HA/β-TCP plasma surface sputtering, which removed the element impurities from the surface, turning the biomaterial into a more advanced material that induced MG-63 proliferation and osteoblastic differentiation. These results are in agreement with our previous studies, in which Ar-GDP treatment combined with fibronectin grafting favored MG-63 cell adhesion, migration, and proliferation on titanium surfaces, suggesting that Ar-GDP treatment improves surface properties [37]. In addition, Mwale et al. [30] modified polymer substrate surfaces using glow discharge treatment and assessed the sensitivity of mesenchymal stem cells to subtle differences in surface chemistry. Their findings revealed that surface modifications regulate osteogenesis and modified mesenchymal stem cell differentiation.

The most common synthetic alloplastic biomaterials in current use are calcium phosphate-based (Ca-P) bioceramics [38], and these have been extensively investigated because their mineral chemistry resembles that of human bone. The present HA/β-TCP has been used successfully for guided bone regeneration in multiple dental and orthopedics treatments [39–41] and has been shown to safely and efficiently support dental implants [42]. These outcomes could be improved in HA/β-TCP with Ar-GDP surface treatments that decrease total surface contaminant contents and prepare stoichiometric HA/β-TCP surfaces in a highly controlled manner. Moreover, Ar-GDP sputtering on HA/β-TCP can be used without modifying the chemical composition of entire particles, as indicated by the absence of differences between HA/β-TCP + Ar-GDP and HA/β-TCP in surface characterizations and XPS analyses.

To confirm the results in the present study, in vivo testing in suitable animal models will be important for assessing HA/β-TCP + Ar-GDP biocompatibility, wound interaction, and efficacy in bone regeneration.

5. Conclusions

Within the limitations of this in vitro study, the present data show that GDP surface treatment only modifies HA/β-TCP chemical surfaces by eliminating contaminants and improving Ca/P mole ratios to values that more closely resemble human bone. Taken together, these experiments warrant further consideration of HA/β-TCP + Ar-GDP as a technology that improves new bone regeneration by enhancing cell attachment, proliferation, and differentiation into osteoblast like cells.

Supplementary Materials: The following are available online at www.mdpi.com/1996-1944/10/12/1347/s1. Figure S1: Fourier Transform Infrared Spectrum (FTIR) Analysis. (A) HA/β-TCP particles before and (B) HA/β-TCP particles after Ar-GDP treatment. These show no change in the spectra as a result of the Ar-GDP treatment.

Author Contributions: Eisner Salamanca and Wei-Jen Chang conceived and designed the experiments; Yu-Hwa Pan and Pei-Ying Lin performed the experiments; Aileen I. Tsai and Peter D. Wang analyzed the data; Ching-Kai Lin, Nai-Chia Teng and Haw-Ming Huang contributed reagents/materials/analysis tools; Eisner Salamanca, Peter D. Wang and Wei Jen Chang wrote the paper.

Conflicts of Interest: The authors declare no conflict of interest.

Funding Sources: The authors are grateful to Taipei Medical University TMU-105-AE1-B10 for research grant support.

References

1. Susin, C.; Wikesjö, U.M. Regenerative periodontal therapy: 30 years of lessons learned and unlearned. *Periodontology 2000* **2013**, *62*, 232–242. [CrossRef] [PubMed]
2. Trombelli, L. Which reconstructive procedures are effective for treating the periodontal intraosseous defect? *Periodontology 2000* **2005**, *37*, 88–105. [CrossRef] [PubMed]
3. Ramseier, C.A.; Rasperini, G.; Batia, S.; Giannobile, W.V. Advanced reconstructive technologies for periodontal tissue repair. *Periodontology 2000* **2012**, *59*, 185–202. [CrossRef] [PubMed]
4. Weiss, P.; Layrolle, P.; Clergeau, L.P.; Enckel, B.; Pilet, P.; Amouriq, Y.; Daculsi, G.; Giumelli, B. The safety and efficacy of an injectable bone substitute in dental sockets demonstrated in a human clinical trial. *Biomaterials* **2007**, *28*, 3295–3305. [CrossRef] [PubMed]
5. Scarano, A.; Piattelli, A.; Perrotti, V.; Manzon, L.; Iezzi, G. Maxillary sinus augmentation in humans using cortical porcine bone: A histological and histomorphometrical evaluation after 4 and 6 months. *Clin. Implant Dent. Relat. Res.* **2011**, *13*, 13–18. [CrossRef] [PubMed]
6. Reynolds, M.A.; Aichelmann-Reidy, M.E.; Branch-Mays, G.L.; Gunsolley, J.C. The efficacy of bone replacement grafts in the treatment of periodontal osseous defects. A systematic review. *Ann. Periodontol.* **2003**, *8*, 227–265. [CrossRef] [PubMed]
7. Orsini, G.; Scarano, A.; Piattelli, M.; Piccirilli, M.; Caputi, S.; Piattelli, A. Histologic and ultrastructural analysis of regenerated bone in maxillary sinus augmentation using a porcine bone–derived biomaterial. *J. Periodontol.* **2006**, *77*, 1984–1990. [CrossRef] [PubMed]

8. Ramírez-Fernández, M.P.; Calvo-Guirado, J.L.; Delgado-Ruiz, R.A.; Maté-Sánchez del Val, J.E.; Vicente-Ortega, V.; Meseguer-Olmos, L. Bone response to hydroxyapatites with open porosity of animal origin (porcine [OsteoBiol® mp3] and bovine [Endobon®]): A radiological and histomorphometric study. *Clin. Oral Implants Res.* **2011**, *22*, 767–773. [CrossRef] [PubMed]
9. Trombelli, L.; Heitz-Mayfield, L.J.; Needleman, I.; Moles, D.; Scabbia, A. A systematic review of graft materials and biological agents for periodontal intraosseous defects. *J. Clin. Periodontol.* **2002**, *29*, 117–135. [CrossRef] [PubMed]
10. Nery, E.B.; Legeros, R.Z.; Lynch, K.L.; Lee, K. Tissue response to biphasic calcium phosphate ceramic with different ratios of HA/βTCP in periodontal osseous defects. *J. Periodontol.* **1992**, *63*, 729–735. [CrossRef] [PubMed]
11. Tamimi, F.M.; Torres, J.; Tresguerres, I.; Clemente, C.; López-Cabarcos, E.; Blanco, L.J. Bone augmentation in rabbit calvariae: Comparative study between Bio-Oss® and a novel β-TCP/DCPD granulate. *J. Clin. Periodontol.* **2006**, *33*, 922–928. [CrossRef] [PubMed]
12. Blokhuis, T.J.; Termaat, M.F.; den Boer, F.C.; Patka, P.; Bakker, F.C.; Henk, J.T.M. Properties of calcium phosphate ceramics in relation to their in vivo behavior. *J. Trauma Acute Care Surg.* **2000**, *48*, 179–186. [CrossRef]
13. Garrido, C.A.; Lobo, S.E.; Turíbio, F.M.; LeGeros, R.Z. Biphasic calcium phosphate bioceramics for orthopaedic reconstructions: Clinical outcomes. *Int. J. Biomater.* **2011**, *2011*, 129727. [CrossRef] [PubMed]
14. Habibovic, P.; Kruyt, M.C.; Juhl, M.V.; Clyens, S.; Martinetti, R.; Dolcini, L.; Theilgaard, N.; van Blitterswijk, C.A. Comparative in vivo study of six hydroxyapatite-based bone graft substitutes. *J. Orthop. Res.* **2008**, *26*, 1363–1370. [CrossRef] [PubMed]
15. Huffer, W.E.; Benedict, J.J.; Turner, A.S.; Briest, A.; Rettenmaier, R.; Springer, M.; Walboomers, X.F. Repair of sheep long bone cortical defects filled with COLLOSS®, COLLOSS® E, OSSAPLAST®, and fresh iliac crest autograft. *J. Biomed. Mater. Res. B Appl. Biomater.* **2007**, *82*, 460–470. [CrossRef] [PubMed]
16. Aronsson, B.-O. Preparation and Characterization of Glow Discharge Modified Titanium Surfaces. Ph.D. Thesis, Chalmers University of Technology, Gothenburg, Sweden, 1995.
17. Aronsson, B.O.; Lausmaa, J.; Kasemo, B. Glow discharge plasma treatment for surface cleaning and modification of metallic biomaterials. *J. Biomed. Mater. Res. A* **1997**, *35*, 49–73. [CrossRef]
18. Gombotz, W.; Hoffman, A. Gas-discharge techniques for biomaterial modification. *CRC Crit. Rev. Biocompat.* **1987**, *4*, 1–42.
19. Kasemo, B.; Lausmaa, J. Biomaterial and implant surfaces: On the role of cleanliness, contamination, and preparation procedures. *J. Biomed. Mater. Res. A* **1988**, *22*, 145–158. [CrossRef]
20. Smith, D.; Pilliar, R.; Metson, J.; McIntyre, N. Dental implant materials. II. Preparative procedures and surface spectroscopic studies. *J. Biomed. Mater. Res.* **1991**, *25*, 1069–1084. [CrossRef] [PubMed]
21. Zhecheva, A.; Sha, W.; Malinov, S.; Long, A. Enhancing the microstructure and properties of titanium alloys through nitriding and other surface engineering methods. *Surf. Coat. Technol.* **2005**, *200*, 2192–2207. [CrossRef]
22. Desmet, T.; Morent, R.; Geyter, N.D.; Leys, C.; Schacht, E.; Dubruel, P. Nonthermal plasma technology as a versatile strategy for polymeric biomaterials surface modification: A review. *Biomacromolecules* **2009**, *10*, 2351–2378. [CrossRef] [PubMed]
23. Wu, S.; Liu, X.; Yeung, A.; Yeung, K.W.; Kao, R.Y.T.; Wu, G.; Hu, T.; Xu, Z.; Chu, P.K. Plasma-modified biomaterials for self-antimicrobial applications. *ACS Appl. Mater. Interfaces* **2011**, *3*, 2851–2860. [CrossRef] [PubMed]
24. Ebrahimi, M.; Botelho, M.G.; Dorozhkin, S.V. Biphasic calcium phosphates bioceramics (HA/TCP): Concept, physicochemical properties and the impact of standardization of study protocols in biomaterials research. *Mater. Sci. Eng. C* **2017**, *71*, 1293–1312. [CrossRef] [PubMed]
25. Zhao, N.; Wang, Y.; Qin, L.; Guo, Z.; Li, D. Effect of composition and macropore percentage on mechanical and in vitro cell proliferation and differentiation properties of 3D printed HA/β-TCP scaffolds. *RSC Adv.* **2017**, *7*, 43186–43196. [CrossRef]
26. *Biological Evaluation of Medical Devices–Part 5: Tests for In Vitro Cytotoxicity*; ISO 10993-5:2009; International Organization for Standardization: Geneva, Switzerland, 2009.

27. Shibata, Y.; Hosaka, M.; Kawai, H.; Miyazaki, T. Glow discharge plasma treatment of titanium plates enhances adhesion of osteoblast-like cells to the plates through the integrin-mediated mechanism. *Int. J. Oral Maxillofac. Implants* **2002**, *17*, 771–777. [PubMed]
28. Huang, H.M.; Hsieh, S.C.; Teng, N.C.; Feng, S.W.; Ou, K.L.; Chang, W.J. Biological surface modification of titanium surfaces using glow discharge plasma. *Med. Biol. Eng. Comput.* **2011**, *49*, 701–706. [CrossRef] [PubMed]
29. Chang, Y.C.; Feng, S.W.; Huang, H.M.; Teng, N.C.; Lin, C.T.; Lin, H.K.; Wang, P.D.; Chang, W.J. Surface analysis of titanium biological modification with glow discharge. *Clin. Implant Dent. Relat. Res.* **2015**, *17*, 469–475. [CrossRef] [PubMed]
30. Mwale, F.; Wang, H.T.; Nelea, V.; Luo, L.; Antoniou, J.; Wertheimer, M.R. The effect of glow discharge plasma surface modification of polymers on the osteogenic differentiation of committed human mesenchymal stem cells. *Biomaterials* **2006**, *27*, 2258–2264. [CrossRef] [PubMed]
31. Chim, H.; Ong, J.L.; Schantz, J.T.; Hutmacher, D.W.; Agrawal, C. Efficacy of glow discharge gas plasma treatment as a surface modification process for three-dimensional poly (D,L-lactide) scaffolds. *J. Biomed. Mater. Res. A* **2003**, *65*, 327–335. [CrossRef] [PubMed]
32. Hing, K.; Annaz, B.; Saeed, S.; Revell, P.; Buckland, T. Microporosity enhances bioactivity of synthetic bone graft substitutes. *J. Mater. Sci. Mater. Med.* **2005**, *16*, 467–475. [CrossRef] [PubMed]
33. Raynaud, S.; Champion, E.; Bernache-Assollant, D.; Thomas, P. Calcium phosphate apatites with variable Ca/P atomic ratio I. Synthesis, characterisation and thermal stability of powders. *Biomaterials* **2002**, *23*, 1065–1072. [CrossRef]
34. LeGeros, R.; Lin, S.; Rohanizadeh, R.; Mijares, D.; LeGeros, J. Biphasic calcium phosphate bioceramics: Preparation, properties and applications. *J. Mater. Sci. Mater. Med.* **2003**, *14*, 201–209. [CrossRef] [PubMed]
35. Tzaphlidou, M.; Zaichick, V. Calcium, phosphorus, calcium-phosphorus ratio in rib bone of healthy humans. *Biol. Trace Element Res.* **2003**, *93*, 63–74. [CrossRef]
36. França, R.; Samani, T.D.; Bayade, G.; Yahia, L.H.; Sacher, E. Nanoscale surface characterization of biphasic calcium phosphate, with comparisons to calcium hydroxyapatite and β-tricalcium phosphate bioceramics. *J. Colloid Interface Sci.* **2014**, *420*, 182–188. [CrossRef] [PubMed]
37. Chang, Y.C.; Lee, W.F.; Feng, S.W.; Huang, H.M.; Lin, C.T.; Teng, N.C.; Chang, W.J. In vitro analysis of fibronectin-modified titanium surfaces. *PLoS ONE* **2016**, *11*, e0146219. [CrossRef] [PubMed]
38. Han, T.; Carranza, F., Jr.; Kenney, E. Calcium phosphate ceramics in dentistry: A review of the literature. *J. West. Soc. Periodontol. Periodontal Abstr.* **1984**, *32*, 88–108. [PubMed]
39. Bensaha, T.; El Mjabber, H. Evaluation of new bone formation after sinus augmentation with two different methods. *Int. J. Oral Maxillofac. Surg.* **2016**, *45*, 93–98. [CrossRef] [PubMed]
40. Vallet-Regí, M. Ceramics for medical applications. *J. Chem. Soc. Dalton Trans.* **2001**, 97–108. [CrossRef]
41. Daculsi, G.; Durand, M.; Fabre, T.; Vogt, F.; Uzel, A.P.; Rouvillain, J.L. Development and clinical cases of injectable bone void filler used in orthopaedic. *IRBM* **2012**, *33*, 254–262. [CrossRef]
42. Seong, K.C.; Cho, K.S.; Daculsi, C.; Seris, E.; Guy, D. Eight-year clinical follow-up of sinus grafts with micro-macroporous biphasic calcium phosphate granules. In *Key Engineering Materials*; Trans Tech Publications: Zürich, Switzerland, 2014; pp. 321–324.

© 2017 by the authors. Licensee MDPI, Basel, Switzerland. This article is an open access article distributed under the terms and conditions of the Creative Commons Attribution (CC BY) license (http://creativecommons.org/licenses/by/4.0/).

Review

A Review on the Use of Hydroxyapatite-Carbonaceous Structure Composites in Bone Replacement Materials for Strengthening Purposes

Humair A. Siddiqui [1,2], Kim L. Pickering [1] and Michael R. Mucalo [3,*]

1. School of Engineering, Faculty of Science & Engineering, University of Waikato, Hamilton 3240, New Zealand; ahumair@hotmail.com (H.A.S.); klp@waikato.ac.nz (K.L.P.)
2. Department of Materials Engineering, Faculty of Chemical & Process Engineering, NED University of Engineering & Technology, Karachi 75270, Pakistan
3. School of Science, Faculty of Science & Engineering, University of Waikato, Hamilton 3240, New Zealand
* Correspondence: michael.mucalo@waikato.ac.nz

Received: 31 August 2018; Accepted: 22 September 2018; Published: 24 September 2018

Abstract: Biomedical materials constitute a vast scientific research field, which is devoted to producing medical devices which aid in enhancing human life. In this field, there is an enormous demand for long-lasting implants and bone substitutes that avoid rejection issues whilst providing favourable bioactivity, osteoconductivity and robust mechanical properties. Hydroxyapatite (HAp)-based biomaterials possess a close chemical resemblance to the mineral phase of bone, which give rise to their excellent biocompatibility, so allowing for them to serve the purpose of a bone-substituting and osteoconductive scaffold. The biodegradability of HAp is low (Ksp $\approx 6.62 \times 10^{-126}$) as compared to other calcium phosphates materials, however they are known for their ability to develop bone-like apatite coatings on their surface for enhanced bone bonding. Despite its favourable bone regeneration properties, restrictions on the use of pure HAp ceramics in high load-bearing applications exist due to its inherently low mechanical properties (including low strength and fracture toughness, and poor wear resistance). Recent innovations in the field of bio-composites and nanoscience have reignited the investigation of utilising different carbonaceous materials for enhancing the mechanical properties of composites, including HAp-based bio-composites. Researchers have preferred carbonaceous materials with hydroxyapatite due to their inherent biocompatibility and good structural properties. It has been demonstrated that different structures of carbonaceous material can be used to improve the fracture toughness of HAp, as they can easily serve the purpose of being a second phase reinforcement, with the resulting composite still being a biocompatible material. Nanostructured carbonaceous structures, especially those in the form of fibres and sheets, were found to be very effective in increasing the fracture toughness values of HAp. Minor addition of CNTs (3 wt.%) has resulted in a more than 200% increase in fracture toughness of hydroxyapatite-nanorods/CNTs made using spark plasma sintering. This paper presents a current review of the research field of using different carbonaceous materials composited with hydroxyapatite with the intent being to produce high performance biomedically targeted materials.

Keywords: hydroxyapatite; carbon; graphene; strengthening; toughening; fracture; crack bridging; nanotechnology; fracture mechanics

1. Introduction

Significant numbers of people around the world have suffered from bone defects that are mainly due to trauma, bone-related diseases, and improper bone tissue growth. This situation has worsened during the recent era primarily because of a burgeoning global population suffering from a proportionately greater number of bone-related diseases, like age-related bone loss [1]. In addition, there are the effects of increasing sports-related injuries and traffic accidents, which have heightened the demand for bone tissue replacement [2]. At present, treatments for bone defects/substitutes include autografts and allografts, both having some important limitations. Limitations that are associated with autografts are mostly linked with a shortage of material available for autografting (i.e., from the iliac crest) and are also (rarely) linked with donor site morbidity. Disease transmission risks and poor immune response describe the main limitations associated with allografts. These limitations have provoked greater research efforts in bone tissue engineering with an aim to utilise a synergistic combination of different materials for functional bone regeneration [3]. In a typical bone tissue engineering protocol, a three-dimensional (3D) porous scaffold is initially made and is then loaded with specific living cells and/or tissue-inducing growth factors to initiate and promote tissue regeneration or replacement [2,4]. The initiative of bone tissue engineering has escalated research in the field of biomedical sciences, i.e., it has resulted in a more biologically focused and coherent way of designing and developing 3D scaffolds with an appropriate/desired porosity so that they can serve as reinforcement, support, and in some special cases, firmly establish tissue regeneration and replacement. A perfect scaffold is one that has an interconnected porous structure to guide new tissue in-growth and regeneration [2]. Large numbers of people, every year, find themselves in need of different kinds of biomaterials, like that needed for dental filling material and for hip joint replacements. After the procedure of blood transfusion, bone grafting has turned out to be the second most recurrently performed clinical procedure each year. However, the harsh reality is that many patients still accept the amputation of diseased/damaged bone/organ as an ultimate treatment, because of the unavailability of suitable bone graft substitute materials [5,6].

At present, various biomaterials are designed and fabricated using polymers, metals, ceramics, or their composites. Bioceramics and their composites have increasingly become an established class of materials applied as human body implants in the form of 3D scaffolds, as they have the necessary properties for biological activity in regard to cell adhesion, migration, and proliferation [2,4,7]. Amongst the different types of bioceramics available, those having a similar chemical identity to that of bone (i.e., calcium phosphate-based ceramics, like hydroxyapatite) have been found to be the most successful, however, their inherently low fracture toughness and strength have historically hampered their use in load-bearing applications [8–12].

This review intends to explore the ways in which the historically poor strength of pure hydroxyapatite implants has been improved through the use of different carbonaceous structures that have the potential to greatly enhance the strength of hydroxyapatite-based composites, manufactured either in the form of a bulk scaffold or as a coating. The article will describe the efforts that have been put together to present the abilities of different carbonaceous structures, especially those in nanoform, to favour crack bridging and deflection in relation to strengthening and toughening effects in a HAp matrix. The biomimetic roles of HAp and carbonaceous structures are also discussed in relation to developing a sound bio-mechanically active implant interface. These roles of carbonaceous materials provide for potential uses in many different biomedical related applications.

Historical Perspective

In the past decades, scientists have focused their research efforts on designing and developing artificial bone, with the objective of creating functional bone regeneration candidates. The historical development of the various stages ("generations") of biomaterials that resulted can be analysed through a consideration of their relative biomedical and mechanical properties. The first generation of biomaterials was mostly related with the development of high strength bioinert materials, which included the development of metallic materials (such as stainless steel and titanium alloys), ceramic materials (like alumina, zirconia), and polymeric materials (like polyethylene and silicone rubber). These biomaterials were used to make devices, which are commonly known as prostheses. At that time, scientists were only focused on an objective to achieve an appropriate combination of physical properties to match those of the substituted tissue (i.e., the defective site) with a minimal toxic response in the host [13]. Almost all of the biomaterials of the first generation had a common attribute, i.e., they were bioinert, and hence not able to develop a biological bond to bone or promote repair of bone. Moreover, they were non-degradable in vivo, which meant that secondary surgery was required if the implant failed and/or needed revision [6,7]. By the end of the 1980's, approximately 50 types of implanted prostheses were in clinical use, which were made using 40 different biomaterials. Many people at that time had had successful surgeries, which saw their lives enhanced due to the use of implants made from 'bioinert' biomaterials. However, it was found (specifically in hip and knee implants) that over time there was a tendency for bone to become slowly resorbed at the prosthesis-bone interface, owing to a phenomenon that is known as "bulk modulus mismatch". This is also known commonly as stress shielding, i.e., almost all of the mechanical load is carried by the higher modulus implant device, which initiates the localised osteoporosis-like symptoms leading to gradual bone resorption, deterioration of bone strength at the implant-bone interface, hence leading to ultimate failure via loosening. At that time, prominent materials scientists of the 1980s, such as Bonfield, began working on the concept of designing biomaterials that would diminish bone resorption due to stress shielding at the prosthesis-bone interface [14].

Later stages of research into biomaterials development were largely associated with advances in the field of artificial bone materials, which moved towards the use of bioactive or biodegradable materials. This era came to be known as that of the second-generation biomaterials. Bioactive ceramic materials (e.g., hydroxyapatite) and biodegradable polymeric materials (e.g., polylactic acid and polyglycolic acid) are important materials comprising the second-generation biomaterials group. Unlike the previous generation of materials, second-generation biomaterials possessed the property to develop strong bonds to surrounding bone tissue and they were biodegradable/bioresorbable. However, both generations of materials were clinically found to be merely functional/structural substitutes i.e., could only bear the load (in the case of the bioinert materials) or (in the case of the osteoconductive materials) provide a scaffold or "structure" in which bone formation had to be induced by natural bodily processes rather than by the implant itself. Some were also found to be insensitive to physiological changes in vivo, such as in the case of stress shielding by bioinert implants [5,11,14–16].

With the inception of multidisciplinary developments in the fields of molecular biology, materials science and manufacturing engineering, many innovative and advanced artificial substitutes having unique functionalities were developed, which are capable of bone repair/regeneration. This has led up to the stage where an important area of research known as tissue engineering was created [17–20]. Tissue Engineering develops scaffolds for placement in defect sites, which not only provide mechanical integrity but also encourage/actively induce stem cell proliferation. A typical procedure of tissue engineering consists of manufacturing a scaffold using suitable biomaterials and subsequently subjecting them to seeding/culturing of stem cells (already harvested from the patient) after which the scaffold is implanted. It can be applied in situations where a temporary functional repair is required or for permanent repair/regeneration. For a successful scaffold material, it is necessary to have an appropriate degradation rate, as very slow rates of degradation retard bone growth, while overly fast rates will result in loss of mechanical integrity. Therefore, it is necessary for a bone scaffold to possess the desired biological, mechanical, and topological properties, as well as favourable osteoconductive properties, thereby facilitating bone remodelling from artificial bone into natural bone [5–7,14–16,21].

In general, calcium, hydroxyapatite (HAp) based bioceramics show excellent biocompatibility, corrosion resistance, and very good compressive strength, which have made them a suitable candidate for implants. However, amongst the list of different calcium phosphates used biomedically, HAp has attracted the most interest from the scientific community due mainly to its features of promoting osseointegration and new bone formation processes, as well as its low toxicity, and resemblance to mineral bone. Research studies so far suggest that, among the list of calcium phosphate compounds, HAp-based materials have been the most successful in bone grafting applications.

2. Hydroxyapatite

The enhancement of advanced materials for biomedical applications is a critical issue challenging modern-day materials science and engineering, especially when it comes to the development of materials to be used in vivo. Historically, hydroxyapatite (HAp), $Ca_{10}(PO_4)_6(OH)_2$ has been the best bone substitute, because it can achieve a sound, firm bond with bone tissue, and furthermore, it can demonstrate osteoconductive behaviour and have no undesirable effects on the human body [15,22–26]. The term "hydroxyapatite" implies the presence of hydroxyl (OH) groups while "apatite" (derived from the Greek word "apatos") meaning "to deceive" was historically used to identify mineral apatites, as they were often mistaken for precious gems, like topaz) [27]. Apatites, in general, are known by the chemical formula $M_{10}(XO_4)_6Z_2$, in which M^{2+} is a metal and XO_4^{3-} and Z^- are anions. In the unit cell of HAp, the M^{2+} is Ca^{2+}, XO_4^{3-} is PO_4^{3-} and Z^- is OH^-. The Ca:P mole ratio in HAp is 1.67. HAp crystallises in a hexagonal system with crystallographic parameters: a = 9.418 Å, c = 6.881 Å, and β = 120°. The crystalline network of HAp is a compact assembly of tetrahedral PO_4 groups, in which phosphorus atoms are found in the centre of tetrahedra having four oxygen atoms at the top. Each PO_4 tetrahedron delineates two types of distinct channels. The first channel is surrounded by Ca^{2+} ions, denoted as Ca(I) (4 per unit cell). The second type of channel contains six other Ca^{2+} ions, denoted as Ca(II). These channels host OH^- groups along the c axis to balance out the positive charge [28].

As long ago as 1926, diffraction and chemical studies of teeth and bone revealed that their inorganic phases were basically calcium hydroxyapatite [29]. In bone, the mineral phase consists broadly of hydroxyapatite but there is also a variety of impurity ions present in the HAp lattice, such as carbonate, magnesium, and sodium ions. Carbonate is one of the most abundant impurity ions with its content being about 4–8 wt.%. This is the reason that bone/hard tissue can be regarded as carbonate-substituted HAp (CHAp) [30,31]. Figure 1 represents the micron and sub-micron features of bone tissue, in which hydroxyapatite platelets can be seen. The biological behaviour of HAp-based ceramics relies on many factors, such as chemical and phase transformation, microstructure, pore size, and pore volume. In surgery, usage of both porous, as well as dense bioceramics, is common, as it depends upon the function and level of implantation that is required by the patient. It is most commonly dealt with on a case-to-case basis rather than in general. Experimentally porous ceramics have a low strength (although strength is found to be dependent upon the level of porosity), and hence such ceramics are found clinically to be more appropriate for drug delivery or for implantation into low load-bearing tissues (like maxillofacial applications) [22,32,33]. Osteointegration depends heavily on the pores in implants, more specifically, the pore size, volume, and their interconnectivity. It was observed that for bone ingrowth into an implant (successful osseointegration) the minimum pore size should be around 100–135 μm, and there is also found to be a direct relation between porosity/pore interconnectivity with the process of bony ingrowth into the implant, as the porosity/pore interconnectivity increases bone ingrowth, especially fixation processes, which become more operative. Moreover, protein adsorption favours separation of osteogenic cells. Subsequently, the presence of small, submicron pores, specifically those about the size of blood plasma proteins, favours biointegration. Thus, it is preferred to have a bimodal distribution of pore-sizes in bioceramics [20,23,33,34].

Small HAp granules are also of interest biomedically, as they find wide application in maxillofacial surgery and implantable drug delivery systems. There are numerous ways to produce HAp granules, like hydrothermal synthesis and pelletising. It was observed through scientific study that spherical HAp particles tend to promote osteointegration and diminish inflammatory processes [20,35]. Ceramics can also serve as a basis for producing composite materials. There is considerable research effort focused on synthesising HAp-matrix composites that achieve reinforcement of the ceramic materials through the use of fine particles, micro lamellae, or fibres, These can raise the strength and toughness of such materials to the level necessary for hard tissue (bone) replacement implants in more load bearing areas [24,36].

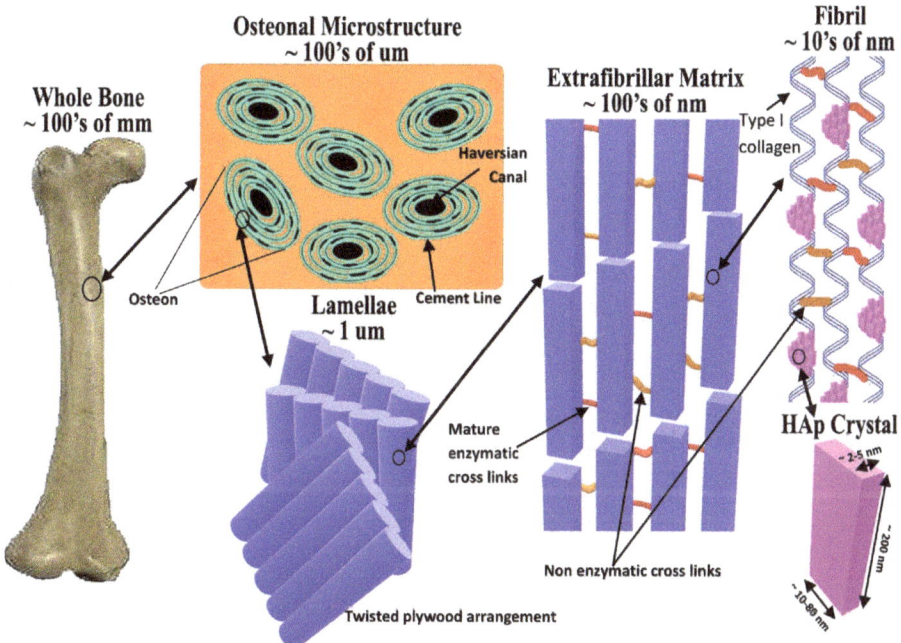

Figure 1. A microstructural representation of bone with size scales.

Researchers have synthesised hydroxyapatite using different techniques, such as hydrothermal synthesis, precipitation, and hydrolysis, as well as employing natural resources, like bovine bones [37–40], fish bones [41–43], marine shells [44–46], and eggshells [45,47–49]). Historically, the usage of xenogeneic bone, like Kiel bone (actual bone harvested from calf or ox) and Boplant (actual bone harvested from a calf) in biomedical applications, were thought to be the alternate to auto- or homografts, mainly due to limited supply of grafts, site morbidity, and the difficulty/expense of multiple surgical processes [22]. A study by Callan and Rohrer [39] also reported patient fears about human tissue transfer, due to the advent of AIDS, especially in dental surgery. They reported that clinicians were actively looking for alternative allograft materials and proposed a naturally derived xenogeneic hydroxyapatite (HAp). The xenogeneic bone grafting material that was used was a fully crystalline naturally porous bovine-derived hydroxyapatite (with all collagenous protein removed), having a particle size in the range of 250–450 μm. Although the organic matter had been previously removed, the bone microstructure was retained. Hydroxyapatite, either synthetic or naturally produced, has been used in wide variety of biomedical applications mainly due to its property of being able to form a bone-like apatite layer at the bone tissue interface [17,50,51]. As mentioned previously, despite these favourable bioactive characteristics of HAp, its poor mechanical properties, like low fracture strength and toughness, preclude its usage for high load bearing applications, such as an implant for the hip or knee. Therefore, to make HAp more suitable for such applications, it needs to be reinforced with various biocompatible materials for improving its strength and toughness. This hence necessitates HAp being prepared as a biocomposite material.

Strengthening and Toughening of Hydroxyapatite—And Its Importance

The understanding and the application of strengthening and toughening mechanisms are very crucial for designing new and optimised materials. A common engineering goal for a structural material is to be both strong and tough, yet often, increase in strength is coupled with a reduction of toughness and vice versa. Strength is simply the material's ability to withstand stress without being subjected to non-recoverable deformation or fracturing, whereas toughness can be regarded as the resistance to crack propagation [52,53]. Every material contains some form of defects (e.g., pores and cracks) and designing with brittle materials, such as alumina (defect-tolerant design philosophy), is challenging as they are influenced by such defects. Fracture mechanics is a valuable tool for design engineers, as it helps in identifying and evaluating conditions under which a crack/defect will propagate to cause a failure, usually by relating three important parameters i.e., the size of the largest/critical flaw, the applied stress, and the fracture toughness (Figure 2). Fracture toughness (or plane strain fracture toughness to use its full title) is a material property that can be obtained through experimental methods, like a three-point bending test, such that crack propagation is instigated. Design engineers can mathematically evaluate critical flaw size or critical stress levels using Equation (1) (fracture toughness equation), once the fracture toughness values are obtained.

$$K_{IC} = Y\sigma\sqrt{\pi a} \qquad (1)$$

where K_{IC} is the fracture toughness, Y is the geometrical factor, σ is the applied stress, and a is the crack length [54].

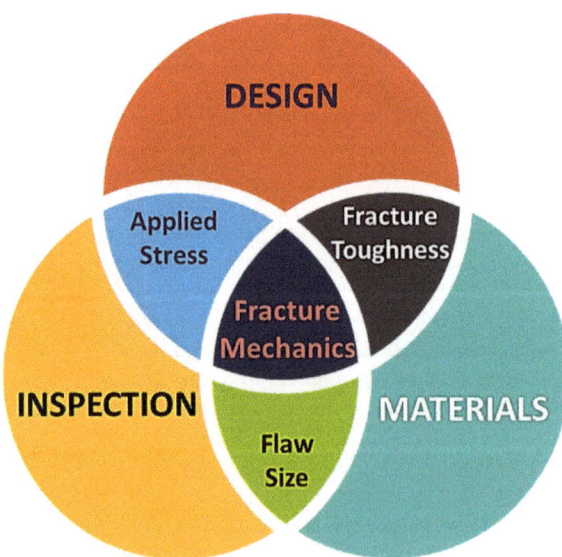

Figure 2. Picture depicting fracture mechanics as an integrated blend of applied stress, flaw size analysis, and fracture toughness.

Ceramic materials are known for their brittleness (lack of ability to plastically deform) [52,54–56]. In fracture mechanics theory (Griffith's theory) fractures inside ceramic materials originate from micro-cracks rather than from atomic bond breaking [57]. Most importantly, these micro-flaws are omnipresent (microscopic flaws, including micro-cracks and internal pores, result from the cooling of the melt or during diffusion-based processes) in the body of a ceramic, and upon loading/stressing these microcracks tend to extend. Similarly, in the case of bioactive ceramics, despite the excellent biological properties, they lack slip systems in their crystals, so whenever they are subjected to an external load, stress relaxing phenomena, like plastic deformation and grain boundary sliding that can occur in metals, do not occur, resulting in poor load-bearing properties for such bioactive ceramics [6]. Hence, general failure mechanisms of ceramic materials are based on the unstable propagation of flaws (pores, cracks, or inclusions), as they are unable to relieve the stress build up at the tip of flaws (through plastic deformation, as, for example, a ductile metal would), and consequently, a ceramic body's strength is dependent upon the combination of the size of the flaw and the applied stress. This explains why they are commonly referred to as notch-sensitive. The flaws in ceramics tend to act as stress concentrating points, i.e., they amplify the stress at the crack tip, which eventually leads to material failure at an applied stress point that is much lower than the theoretical stress [57] i.e., a significant decrease in the values of actual strength when being compared with theoretical values [57]. The theoretical strength of pristine glasses is about 7000 MPa, but practically their strength is merely 1% of the theoretical value (35–70 MPa) [58]. The dimensions (in micron or nanometre), geometry (penny-like round) and orientation (parallel to applied load) of cracks regulate the magnitude of this amplification of stress. This concentration of stress at the crack tip is commonly denoted by a stress concentration factor, which is the ratio between the maximum stress developed at the microstructural defect to the nominal stress in the body.

It is important to highlight the difference between the stress intensity factor and the stress concentration factor. The stress concentration factor, as defined earlier, is the ratio of maximum (which develops at the crack or flaw) to nominal stress in the body and it is dependent upon the geometry of the flaw or crack. Theoretically, the most important factor that is responsible for severe concentration of stress is a sharp crack (zero tip radius). In contrast, the stress intensity factor is used to predict the stress intensity near the crack/flaw tip that was developed due to remote or residual load or simply it evaluates/calculates the local driving force for crack propagation. This factor is dependent together on the geometry of crack and the applied load. The stress intensity factor is calculated for a given geometry and load and compared with a threshold value of stress intensity factor above which cracks will propagate in the given material. This threshold value of the stress intensity factor is known as fracture toughness or critical stress intensity factor. In Equation (1), the product "$\sigma\sqrt{\pi a}$" is the stress intensity factor, which represents how much the applied stress σ, gets intensified at the tip of the crack having a length a, so, technically, in brittle materials, for the fracture to occur, the stress that is developed at the crack tip must be greater than the fracture toughness values [54,55,59]. For bioceramics, where fracture events can be catastrophic, the approach based on utilising fracture toughness values, is the only method that can predict ceramic fracture.

When the stress developed at the crack tip exceeds the fracture toughness value of the material, the crack will fracture. It is important here to discuss the fracture process of this crack first, as the complete fracture is just a coalescence of the propagating crack with other ones. Fracture mechanics defines this process as a competition between two phenomena i.e., intrinsic (damage) processes and extrinsic crack-tip shielding mechanisms. Intrinsic damage processes, such as micro-void coalescence, promote the propagation of a crack tip by operating ahead of it and are dependent upon the nature of microstructure (or nanostructure) or any second phase ahead of crack tip (usually by cracking/debonding), while the extrinsic crack-tip-shielding mechanisms attempt to inhibit/resist this propagation of the crack tip and operate behind the crack tip. One of the principal methods of increasing fracture toughness is by increasing the microstructural resistance, which will enlarge the plastic zone ahead of the crack tip, eventually making initiation as well as propagation of the crack difficult. This is termed intrinsic toughening and is relevant to ductile materials. For ceramics/bioceramics, extrinsic toughening is perhaps the primary source of toughening, as intrinsic toughening would involve altering the bond strength, which is not feasible. Extrinsic toughening is generally based on diverse microstructural mechanisms that play their role (behind the crack tip) in reducing the damaging force developed at the crack tip; this is usually known as crack-tip shielding and can also involve processes like crack bridging by fibres and in situ phase transformation. Intrinsic toughening mechanisms affect crack initiation as well as crack propagation. They are an inherent property of the material and are found to be active in the material irrespective of the crack size and its geometry. Extrinsic mechanisms, which are only effective for crack growth, come in to play in the event of the crack wake and they are dependent on the size and geometry of the crack [52,53,60,61]. In the domain of biofabrication using bioceramics, designers are constantly challenged with the issue of low strength and toughness, for which extrinsic toughening is the only solution that is usually done by reinforcing bioceramics using biocompatible and effective materials [6,17].

Many successful approaches have been undertaken to toughen the ceramic material. The common strengthening and toughening mechanisms are:

1. Modulus/load transfer; i.e., the use of high elastic modulus fibres in a relatively low elastic modulus matrix [62].

2. Pre-stressing; which involves placing a portion of the ceramic under a residual compressive stress.

3. Crack shielding; mainly involving transformation toughening, in which reinforcement particles undergo sudden volumetric change due to an applied stress, which in turn, compresses the flaws in ceramic materials.

4. Crack deflection or impediment; mainly involving the modification in the microstructure of ceramic materials, like dispersing foreign particles, which tend to impede or deflect an advancing crack.

5. Crack bridging; implying the addition of some secondary phase, usually fibrous structures, which bridge the flaws in the ceramic, resulting in enhanced strength.

6. Fibre pull out; is associated with fibre debonding and the corresponding frictional sliding, which enhances the fracture toughness. Some part of the energy is consumed due to friction as fibre, particle, or grain slides against adjacent microstructural features, resulting in enhanced fracture toughness.

Figure 3 represents an opaque and see-through image of the body containing fibrous reinforcement. These pictures clearly outline different strengthening and toughening mechanisms. Figure 4 represents a series of progressive SEM imaging of bovine (rib) bone sintered at high temperatures to burn out all of the collagen protein and lipid residues to yield crystalline hydroxyapatite.

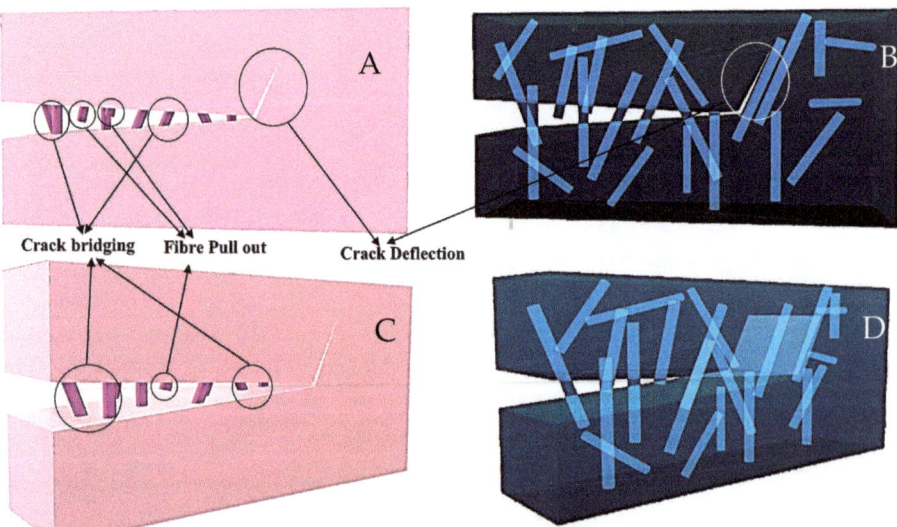

Figure 3. A typical representation of strengthening and toughening by fibres; (**A,C**) represent two views of crack in a body being perturbed by fibres, (**B,D**) represents the see-through image of the body presented in A and C, for a better understanding of the reinforcement effects.

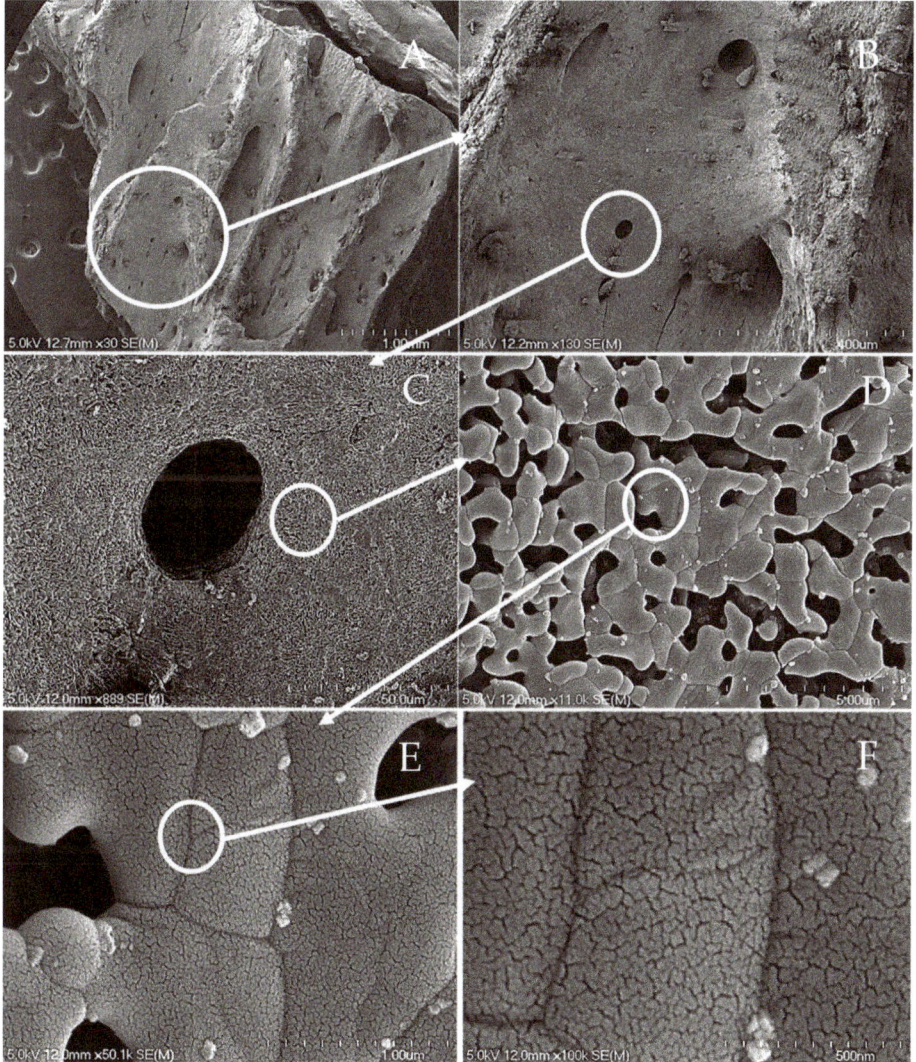

Figure 4. A series of SEM images of surface morphology of high temperature sintered bovine rib bone; (**A**) a particle of bovine bone-derived (CNF)/hydroxyapatite (HAp), (**B**,**C**) presents one of the interconnected pores, (**C**–**E**) presents surface imaging of a HAp particle & (**F**) depicts surface sub-micron cracks on the surface of HAp. These materials were generated by one of the review's authors (Siddiqui).

3. Carbon and Its Structures

Carbon, one of the most abundant elements on earth, is the essential building block of all living organisms. The use of carbon/carbonaceous structures for biomedical applications is a novel trend as new diverse applications are continually being conceived. Carbon has many different allotropic forms and after hydrogen, is the material that can form the most compounds of any of the elements. Some well-known allotropes of carbon are graphite, diamond, glassy carbon, pyrolytic carbon, fullerenes (C_{60}, C_{70}), hexagonal diamond (lonsdaleite), and carbon nanotubes [63]. A broad classification of carbon allotropes is presented in Figure 5, while Figure 6 shows pictures of the different carbonaceous structures possible.

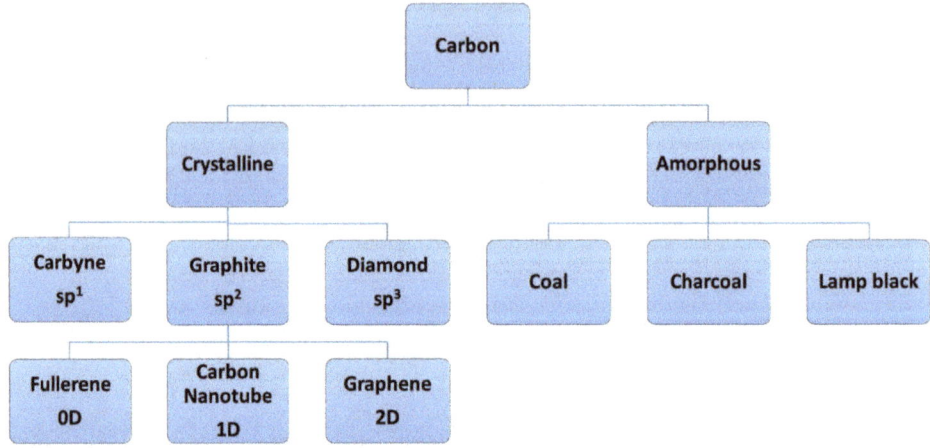

Figure 5. A broad classification of Carbon and its structures (0D, 1D and 2D means zero dimensional, one dimensional and two dimensional, respectively).

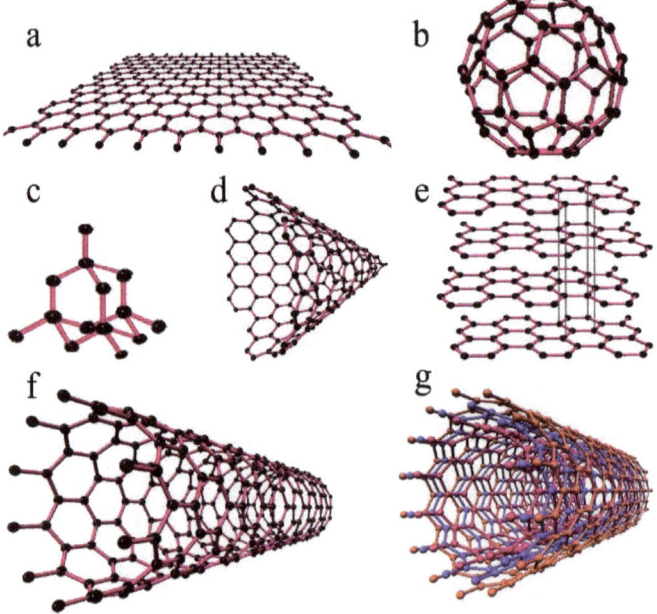

Figure 6. Some important carbonaceous structures, (**a**) Graphene, (**b**) Fullerene C_{60}, (**c**) Diamond, (**d**) NanoCone, (**e**) Graphite, (**f**) Single-wall Carbon Nanotube, and (**g**) Multiwall Carbon Nanotube.

Carbonaceous materials have been finding ever-increasing applications in areas spanning the electronics, mechanics, civil engineering, and medical disciplines. Carbonaceous structures, especially carbon fibres, have a unique set of characteristics, namely, high elasticity modulus, as well as specific thermophysical, electrophysical, and sorption properties. This encourages the ever-expanding introduction of carbon composites into the most topical and knowledge-intensive branches of science and technology [64,65]. The biocompatibility of carbonaceous structures with native tissues of the body invites opportunities for medical applications (creation of artificial heart valves, the fixation of bone fractures, burns and diabetic ulcers, the creation of sorbing elements, and air filters). The creation of composites that are based on HAp and carbonaceous structures is a promising strategy for improving the mechanical characteristics of HAp-based implants [17,51,66–68].

Carbonaceous structures, if used in a limited/low quantity, tend to disperse more evenly and uniformly, leading them to develop good interfacial bonding within the ceramic matrix. It was observed in scientific studies that structures, like graphene and graphene oxide, were effective in enhancing fracture toughness, hardness, and wear resistance of ceramics, which is mainly due to their large surface area and wrinkled surfaces, as such surfaces would give rise to creative mechanical interlocking with the matrix [69]. Carbonaceous structures possess large elastic modulus (900–1000 GPa) values, which are significantly higher than those that were measured for bio-ceramic-constructed forms (40–50 GPa). As a result of this, when a crack propagates in these composite materials, the stress in the matrix can be efficiently transferred to the carbonaceous structures. At this point of crack propagation, carbonaceous structures will be subjected to stretching and fracturing at their ultimate strain values or will be dragged out from the matrix as the stress exceeds the critical interface bonding strength. In any of these scenarios, a significant amount of the fracture energy will hence be consumed. As the crack spreads further, a mechanism of crack bridging may start to appear, due to other carbonaceous structures being present in the matrix. The carbonaceous structures may start to act like "elastic bridges" to impede the further expansion of the crack [59,70]. In addition, to the bridging effect, there can be a crack deflection phenomenon operating, in which the crack interferes with the carbonaceous material, and instead of following its original path of propagation, it is deflected towards the ceramic/carbonaceous material interface. However, if the external energy is inadequate to debond the carbonaceous structure/matrix junction, the crack (crack tip) will become seized, i.e., a condition is achieved, like a shielding effect, which enhances fracture toughness by avoiding further fracturing of the matrix. These above-mentioned mechanisms make further propagation of a crack difficult as most of the external energy is consequently absorbed with little energy remaining for additional spread of the crack. Consequently, the crack spreads in a step-like pattern and progressively decelerates, or even ceases, thus increasing the fracture toughness. This provides a strengthened material capable of load-bearing applications. Therefore, introducing a secondary phase like a carbonaceous structure, which can form a strong interfacial bond with the ceramic materials, can play an effective role in impeding crack propagation and producing more robust and stronger bioceramics.

Optimisation experiments have indicated that the quantity of carbonaceous material is very important for mechanical properties, as when the content level approaches a certain threshold value, the optimum mechanical properties can be achieved, however, beyond that threshold value, mechanical properties begin to deteriorate [71]. This is attributed to the unwanted agglomeration of carbonaceous structures at high content levels, due to strong van der Waals forces, p-p stacking, and the deficiency of functional sites in carbonaceous structures. Apart from that, dispersion in a ceramic matrix is also inherently difficult, which can result in defects and weak interface bonding, eventually leading to a poor reinforcing effect. Hence, for achieving optimum properties of ceramic-based composites, optimisation of content, as well as the dispersion of the reinforcing phase, are all important requirements. So far, well dispersed carbonaceous structures/ceramic composite powders have been attained either by physical mixing or by colloidal processing. In a typical physical mixing, carbonaceous structures were initially ultrasonicated and then mixed with the ceramic powder by milling (usually ball milling) in a solvent. Ultrasonication of carbonaceous structure ensures disaggregation, which makes it suitable for mixing. In a typical colloidal processing protocol, both carbonaceous material and ceramic particles are modified in a way to induce opposite charges on their surfaces, due to which carbonaceous structures can then easily and effectively be dispersed in the ceramic powder via electrostatic attraction, albeit with ultrasonication and stirring being needed to effect efficient mixing. These modifications are brought about usually by surface functionalisation via acidification and oxidation or by using surfactants. As a result, through appropriate processing techniques, carbonaceous structures can mechanically strengthen a hydroxyapatite matrix via crack bridging and deflection mechanisms [6,57,72].

3.1. Pyrolytic Carbon

Pyrolytic carbon is highly fatigue resistant and it possesses elastic modulus values that are similar to that of bone. Historically, pyrolytic carbon was widely used for implant coating/construction, as such coatings tended to resist blood clotting on their surface and to make a frictionless surface, which is very useful in cases where the implant needs to have some sort of relative motion with other implants/surfaces. This material was also used for coating orthopaedic implants, like those that are used for finger joint implants, but was limited in their use in orthopaedic implants due to poor bone bonding properties [73,74]. Hetherington et al. [75] tried to investigate the bone response to HAp-coated pyrolytic carbon, as they plasma-sprayed HAp on pyrolytic carbon implants. Cylindrical samples of pyrolytic carbon, HAp-coated pyrolytic carbon, and HAp-coated titanium were implanted in the femurs of beagle dogs. After eight weeks, the beagles were euthanized, and the implants excised and investigated, which led to the discovery that the pyrolytic carbon had almost nil attachment strength (1.59 MPa), while the other two showed similar interfacial strengths (~8.71 MPa). This study showed that HAp favours a durable bond with the adjacent tissue. Similarly, Lin and Jiarui [73] have reported HAp coating on pyrolytic carbon via electrophoretic deposition (EPD) in glycol and ethanol separately (as the dispersion medium), as they were studying the effect of dispersing media on coating quality. However, this coating was characterised for its chemical and biological properties only, and not for its mechanical properties.

3.2. Carbon Fibres

Carbon fibres (CF) are crystalline filaments of carbon, having a regular hexagonal pattern of carbon sheets. Carbon fibres are widely used as a reinforcing material in polymer and ceramic matrices due to their high elasticity modulus, high strength-to-weight ratio, sorption, and thermophysical properties. The biomedical properties of CFs make them a very important material, as due to their inherent biocompatibility (both in vitro and in vivo), they have been used in making artificial heart valves, in treating bone fractures and purulent wounds, and in making biocomposites. They are usually manufactured via high-temperature conversions during the pyrolysis of carbon-rich (mostly polymeric) precursors [76].

N. A. Zakharova et al. [77] reported the successful manufacturing of composites comprising micrometre-sized carbon fibres (CFs) and biocompatible nanocrystalline calcium hydroxyapatite, containing 1.0, 2.0, and 5.0 wt.% CFs. This study was based on the principle of co-precipitation of HAp-carbon fibre with hydroxyapatite from a solution of calcium and orthophosphate ions and it was noted that the presence of carbon fibres noticeably affected hydroxyapatite crystallization from a system comprising $Ca(OH)_2$–H_3PO_4–CF–H_2O. Researchers found a progressive decrease in HAp sizes in response to increasing carbon fibre percentage, mainly because the presence of more carbon fibres meant more nucleation sites available, which indirectly lead to the formation of finer sized HAp particles. Slosarczyk et al. [78] manufactured a HAp—carbon fibre composite by hot pressing (using a temperature of 1100 °C, pressure of 25 MPa and an argon atmosphere). The composite was found to possess enhanced fracture strength and toughness. It was reported from this study that the effect of hot pressing and carbon fibre presence both played a substantial role in the strengthening and toughening effects that were observed.

Dorner-Reisel et al. [79] studied the microabrasion resistance of non-reinforced and CF-reinforced hydroxyapatite. The composite was made using commercial grade HAp and CF via hot pressing (using a pressure of 25 MPa, temperature of 1000 to 1150 °C, pressing time of 15 min, and an argon atmosphere). Apart from the chemical and structural characterisation, a ball crater test was utilised to ascertain the microabrasion resistance of the composite. It was observed that 20 vol.% of added CF increased the resistance of the composite against microabrasion. Cracking at the interface of the carbon fibres and the HAp matrix was observed after a wear test due to thermal mismatch. To prevent this, the investigators coated the CF with pyrolytic carbon.

Very recently, Boehm et al. [80] tried to investigate the full potential of reinforcing with carbon fibre by undertaking surface functionalisation of the CF. They worked on the concept of improving the wettability (which helps in precipitation) of a carbon fibre by activating their surface, which eventually leads to better adhesion of the fibre with the matrix. Fibres were modified using aqua regia and then calcium adhesion was performed while using $CaCl_2$, which proved influential for the precipitation of calcium phosphate crystals on the fibre surface. This type of fibre modification process has great potential for making a strong composite for load-bearing applications.

3.3. Nano Carbonaceous Materials

Nano carbonaceous materials involve the use of several different types of materials with varied appearance and properties. Some important materials, for example, are carbon nanotubes, graphene, graphene oxide, and buckyballs. All these materials exhibit unusual electrical, mechanical, optical, thermal, and magnetic properties as compared to their bulk form, mainly due to quantum effects and large surface-to-volume ratios. Due to such unique properties, they are currently being used in a number of applications, such as in sensors and smart materials. When employed as second-phase materials for composites, nanotubes and sheet-like structures of carbon are successful due mainly to their large specific surface areas [69].

3.3.1. Carbon Nanotubes

Carbon nanotubes (CNTs) are a type of one-dimensional nano-scaled material, having a tubular structure made by rolling single/multi-layer graphite sheets. CNTs, after their accidental discovery in 1991 by the Japanese physicist, Iijima, have piqued the interest of the scientific community mainly because of their unique structural and physicochemical properties. They possess large specific surface areas (50–1315 $m^2\ g^{-1}$), a high aspect ratio (as they have their diameters at the nano-scale but their lengths, are, by contrast, at the micron-scale), high tensile strength, high resilience, and flexibility. The strength of CNTs is approximated to be roughly 100 times that of steel, but its density is just a fraction of that of steel. The reason for these highly favourable mechanical properties is due to the CAC (Carbon Atomic Chain) covalent bond that exists in the carbon rings, as these bonds are considered to be very stable chemical bonds. On the basis of this physical structure, carbon nanotubes can be single-walled carbon nanotube (SWNT) (single tube)—or multi-walled carbon nanotubes (MWNTs) (concentric cylinders of carbon) [71,81–83].

There are many different processes to prepare carbon nanotubes, however, arc-discharge [84,85], laser ablation [81,84,86,87], and chemical vapour deposition (CVD) [85,88,89] are the most common. For large quantities on a commercial scale, CVD is widely used for the preparation of CNTs. In a typical CVD process, the hydrocarbon feedstock is reacted with a metal catalyst at elevated temperatures to yield carbon nanotubes, however, the quality, quantity and dimensions of CNTs are heavily dependent on the reaction conditions. Owing to their unique properties, especially their property to act as carriers for bone morphogenetic proteins, they are regarded as highly successful candidates for reinforcing hydroxyapatite and numerous scientific studies have been performed to ascertain their strengthening and toughening effects in bio-ceramics. Most biomedical research using such composites shows that CNT-based biocomposites are appropriate for cell growth and enzyme activity [71,82].

Chen et al. [90] developed a strong CNTs-reinforced hydroxyapatite composite coating on titanium (Ti-6Al-4V) alloy. The composite coating was made while using a sophisticated technique known as "laser surface alloying". Initially, commercial grade HAp and CNTs were mechanically ball milled in varying percentages of CNTs (0%, 5%, 10%, and 20%). A Nd: YAG laser, (output power 400 W) was used to apply the coating. After mechanical characterisation a significant increase in hardness was observed but only a slight improvement in modulus. In general, it was found that approx.

7 vol.% of carbon nanotubes increased fracture toughness by 50% and flexural strength by 28%. In comparison, Lahiri et al. [91] reported an increase of 92% and 25% in fracture toughness and elastic modulus values with the addition of 4 wt.% CNTs. Lahiri presented an increase of 260% in fracture toughness and of 50% in flexural strength in hydroxyapatite-nanorods/CNTs by the addition of 3 wt.% CNTs manufactured using spark plasma sintering. It was also reported that vacuum hot-pressing of hydroxyapatite ceramics also improved the mechanical properties [92].

Kealley et al. [93] developed a HAP-CNT composite via a chemical precipitation technique. Commercial CNTs (2 wt.%) were added during the precipitation stage of HAp manufacturing. The powder that was obtained was subjected to hot isostatic pressing at 900 °C under an argon atmosphere. In situ neutron diffraction was utilized to confirm that CNTs and hydroxide bonds were present in the composite. Kaya [94] developed a CNT-reinforced HAp coating, in which the addition of only 2 wt.% CNTs was found to significantly improve the hardness (i.e., by 10-fold), elastic modulus (by six-fold), and shear strength (by 3-fold) of the coating. Moreover, CNTs were also found to prevent spalling of the coating. CNTs were initially acid-treated to adjust the surface charge, which helped in the dispersion of CNTs in the HAp. Meng et al. [95] utilized nano-sized HAp particles with needle-like morphology to make a HAP-CNT composite via hot pressing. The composite was found to have enhanced mechanical properties. CNTs were initially functionalized to introduce -COOH groups on its surface, which helped in developing strong interactions between these and the Ca ions on the HAp particles. The composite demonstrated an approx. 50% increase in fracture toughness and a 28% increase in flexural strength with the addition of 7 vol.% CNTs. Sarkar et al. [96] utilized surfactant-modified CNT to make HAp-CNT composites via a spark plasma sintering process while using surfactant-modified CNT and HAp nanopowders. HAp was synthesised using a microwave-assisted process and was ball milled with 2.5 vol.% of CNTs. The resultant powder was spark plasma-sintered at various temperatures, where the temperature of 1100 °C produced the maximum fracture toughness of 1.27 MPa.m$^{1/2}$.

Balani et al. [97] developed CNT-reinforced HAp coatings using plasma spraying. The powder feedstock for plasma spraying was comprised of 4% CNTs blended with HAp, which was sprayed on titanium Ti-6Al-4V substrate (power 20–25 kW). The coating was found to be highly crystalline and non-toxic with a ca. 56% improvement in fracture toughness. Li et al. [98] utilised a double in situ process to develop HAp-CNT composites. In the double in situ process, CNTs were initially synthesised and modified using HAp via CVD, and then were further encapsulated using HAp via a sol-gel process. The modification of CNTs with HAp proved to be helpful in developing a strong interfacial bond and a homogenous dispersion in the matrix, which led to enhanced biological as well as mechanical properties, as the flexural strength was found to be higher (1.6 times) than that of pure HAp. Kim et al. [99] reinforced HAp with different percentages of multiwalled carbon nanotubes and utilised the spark plasma sintering (SPS) technique for consolidation. Sintering at 900 °C, the composite attained maximum density, fracture toughness, and Vickers microhardness, but sintering beyond this temperature led to a degradation of the properties. It was also observed that increasing nanotube concentration was directly increasing the hardness and fracture toughness. A pull out mechanism was widely observed and declared to be the origin of the elevated fracture toughness values.

3.3.2. Graphene

Graphene is a leading nanomaterial that commands a strong research following due to its diverse applications and properties. This material has also been of strong interest to biomaterial scientists due to its two-dimensional structure and large contact area, which makes it an ideal material for making composites. It is composed of a single layer of carbon (sp^2-hybridized) atoms arranged in a honeycomb lattice-like arrangement. Graphene is the basic building block of all graphitic materials and possesses exceptional mechanical properties like tensile strength of ca. 130 GPa and a Young's modulus ca. 0.5–1 TPa [100,101]. It has been found that graphene can stimulate a toughening effect in a HAp matrix even at a low content level; however, it would be appropriate to term this reinforcing graphene as "graphene nanosheets (GNS)". This is because GNSs usually have thicknesses near 8–10 nm and are usually made up of a few graphene layers. GNS show similar properties to that of graphene (monolayers), in that they have strong mechanical, high electrical, and enhanced biological properties. From graphene can be derived two further important materials i.e., graphene oxide (GO) and reduced graphene oxide (RGO), which both have unique properties. Many biomaterial scientists have developed graphene-HAp based biocomposites for different biomedical applications, especially those that can be used for orthopaedic applications [102].

Liu et al. [103] developed graphene/hydroxyapatite composites using precipitation methods, as a result of which nucleation and growth of rod-like HAp nanoparticles on the graphene surface was observed, which occur most likely as the result of a charge balancing mechanism. After using arc spark plasma sintering, there was a significant increase in the fracture toughness (203%) with 1.0 wt.% graphene. In contrast, Zhang et al. [104] reported a maximum of 80% increase in the fracture toughness of HAp-1 wt.% CNT composite made via spark plasma sintering. This rise in fracture toughness was attributed to different strengthening and toughening mechanisms, like crack bridging, crack deflection, and pull-out effects. An in vivo study [105] on mice showed that the addition of graphene did not alter biocompatibility. Therefore, it is a very promising strategy to incorporate graphene as a second phase reinforcement in HAp, as it can enhance the mechanical properties of HAp without compromising the biological properties.

3.3.3. Graphene Oxide

Graphene oxide (GO) is an important derivative of graphene. It can be referred to as the oxygenated counterpart of graphene as it consists of graphene sheets covered with oxygen-based functional groups, like hydroxyls (on its planes), epoxide (on its planes), and carboxyl and carboxyl groups (on its edges). Its particularly large surface area, biostability, biocompatibility, antibacterial properties, ease of chemical functionalisation, favourable dispersion behaviour, and mechanical properties are the main reasons for its widespread applications in diverse fields. The presence of functional groups on its surface facilitates in achieving dispersion stability and enhancement of interfacial bonding, which eventually leads to better load transfer in composites [106–108]. As compared to CNT, high-quality GO can be manufactured on a commercial scale at a lower cost. All of these properties and its relatively low cost of production have made GO an important material for nanoscale reinforcement in biocomposites. Several studies were conducted to synthesise HAp-GO based composites, as the incorporation of GO in HAp will ensure a good dispersibility and enhancement of mechanical strength. Historically, Schafhaeutl [109] had reported GO for the first time in 1840 (while working carbon for making cast iron, steel and malleable iron) and then Brodie [110] in 1859 (while discussing and evaluating the atomic density of graphite). At present, GO is mostly synthesised using a process proposed in 1958 by Hummers and Offeman [111] in which graphite is initially oxidised (using concentrated sulfuric acid, sodium nitrate, and potassium permanganate) to achieve GO having an increased interlayer distance addition of oxygen-containing functional groups. If this graphite oxide is subjected to ultrasonication, then it will dislodge the graphitic sheets to achieve graphene oxide (GO). There is also a modified Hummers method in which $NaNO_3$ was eliminated to make the process safer and environmentally friendly, but the main strategy is the same. GO can also be reduced to achieve graphene, but due to partial reduction instead of graphene, "reduced graphene oxide" (RGO) usually results. Figure 7 depicts how GO and RGO can be obtained from graphite.

Li et al. [112] reported a significant reduction in surface cracks and an improvement in resistance to coating detachment in the HAp based coating by adding graphene oxide (GO) via a cathodic electrophoretic deposition process on a titanium substrate. Nano GO and HAp were commercially sourced and mixed together in varying percentages of GO (2% and 5%) via an ultrasonic process to ensure a homogenous distribution. The coating was achieved while using the electrophoretic deposition method with 30 V of constant voltage in a HAp-GO suspension. The coating was found to have superior corrosion resistance and in vitro biocompatibility, along with enhanced coating adhesion properties (ca. 75% increase with 2% GO and ca. 110% increase with 5% GO, when compared to the adhesion strength of pure HAp). Fathyunes and Khalil-Allafi [113] utilized ultrasound-assisted pulse electrodeposition to develop microstructurally refined and compact GO-HAp coatings on titanium substrates. The utilization of ultrasonic power (>60 W) was found to be effective for the incorporation/penetration of GO sheets into the coating. GO was synthesised using the modified Hummers' method. The electrodeposition method was performed using an electrolyte containing $Ca(NO_3)_2$, $NH_4H_2PO_4$, and H_2O_2, with the addition of a 100 µg/mL GO suspension, which resulted in the development of the GO-HAp coating. The complete process of electrodeposition was aided by ultrasonication with the HAp-GO coating being found to have higher hardness and modulus values (3.08 GPa of nano-hardness and 41.26 GPa of elastic modulus at the ultrasonic power of 60 W) together with enhanced corrosion resistance. Li et al. [114] successfully developed a GO-HAp composite using an in situ one-step mineralization method. GO was synthesised using the modified Hummers and Offema method. To perform the in situ mineralisation, GO and calcium chloride were ultrasonically mixed in a mixture of water and ethylene glycol and then disodium hydrogen phosphate solution was added, which stimulated the mineralisation of HAp, as confirmed by transmission electron microscopy (TEM), X-ray diffraction (XRD), Energy dispersive spectrometry (EDS), and Atomic Force Microscopy (AFM). Mechanical testing of GO-HAp and GO sheets/papers revealed higher modulus values (16.9 GPa), higher tensile strength (75.6 MPa), but reduced fracture toughness values (214.9 kJ m^{-3}), as compared to GO alone. The modulus values of GO-HAp were found to be analogous to those measured for the human femur bone (13–15 GPa) and human tibia bone (13–16 GPa).

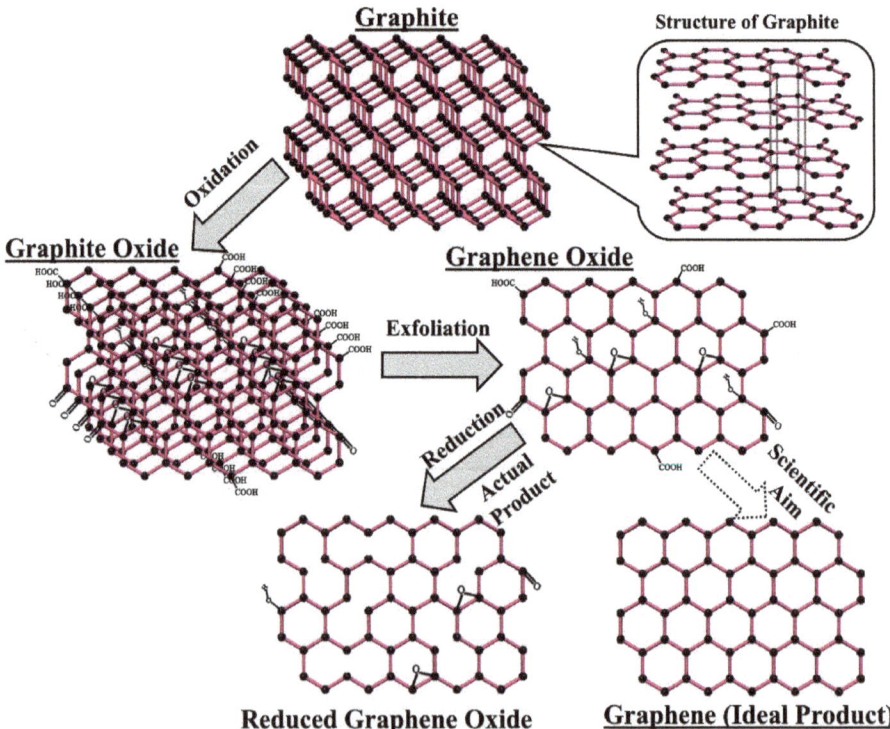

Figure 7. Scheme for making graphene oxide and reduced graphene oxide from graphite.

3.3.4. Reduced Graphene Oxide

One attractive feature of GO is its ability to undergo (partial) reduction by removing oxygen-containing groups to yield graphene-like sheets. This recently developed material is a type of graphene and it has many terms describing it like "functionalized graphene", "reduced graphene", or "chemically modified/converted graphene", but, in general, it is known as "Reduced Graphene Oxide" or RGO. The scientific aim of the reduction process of graphite/graphene oxide is to make pristine graphene, however, there always remain some defects and functional groups (residual) defects that can harshly alter the structure of the carbon plane. The reduction process can be chemical, thermal, or electrochemical. RGOs possess a honeycomb crystal lattice and they are found to have unique properties, like high electrical and thermal conductivity, biocompatibility, high surface area (>2600 m^2 g^{-1}), and chemical stability [108,115].

Liu et al. [103] developed HAp-RGO nanocomposites while using a liquid penetration technique. RGO was synthesised using the modified Hummer's method and it was made into a suspension by dispersing in water. In the RGO suspension, HAp nanorods were wet-chemically synthesised to produce a HAp-RGO composite. HAp-RGO composite powders with varying percentages of RGO were pelletised using Spark Plasma Sintering. The composite's fracture toughness values reached up to 3.94 MPa.m$^{1/2}$; much higher than that for pure HAp. Crack tip shielding, crack deflection, and bridging were considered to be the reasons for the higher observed fracture toughness values, which were brought about by the inclusion of the RGO in the composite.

Baradaran et al. [116] developed HAp-RGO composites, in which HAp was used in the form of nanotubes. GO was prepared while using the simplified Hummer's method, while HAp nanotubes were made using a surfactant-free solvothermal process. HAp-RGO composites were synthesised using a hydrothermal process during which GO was reduced to RGO then formed into the hybrid material. The powder was pelletised using hot isostatic pressing using 160 MPa pressure at 1150 °C. The composite showed an increase in the fracture toughness and elastic modulus of 40% and 86% respectively when compared to pure HAp. Biological testing revealed the promotion of osteoblast and proliferation activity on the composite. Similarly, Elif et al. [117] successfully reinforced HAp with RGO. RGO was made from GO, which was initially made using Hummer's method, and then it was reduced using plant extracts to yield RGO. HAp-RGO composites were made using a liquid penetration method and were subjected to pelletising and sintering. It was observed that a small quantity (1%) of RGO had caused an increase of 3.2 times in the compressive strength of the composite as compared to the equivalent pure HAp sample. An increase in biocompatibility was also observed.

3.3.5. Nanodiamonds

Nanosized diamond particles (Nanodiamond (ND)) are known for their mechanical, biological, and tribological properties, which include biocompatibility, high hardness, a low friction coefficient and chemical stability. This unique set of properties in ND has ignited intense interest for its applications as a secondary phase for reinforcing a bioceramic, especially a HAp matrix. On a commercial scale, NDs are formed by the detonation of carbon-based explosives and they are hence known as detonation nanodiamonds. In a detonation nanodiamond, there can be impurity species like O, N, Fe, Cr, Ca, and some functional groups like N–H, C–O–C, C–OH, C=O present which is given as the reason for the surface's chemical multi-functionality. Nanodiamonds have also been used to produce mechanically strong HAp based coatings for load-bearing biomedical applications. The addition of nanodiamonds in HAp based coatings may also increase coating adhesion and prevent metal ion release from metal surfaces. Nanodiamonds have also been explored for their application in drug delivery applications [118–120].

Pramatarova et al. [121] developed and studied biomedical coatings that were based on hydroxyapatite reinforced with detonation nanodiamonds (DND). The coating was grown biomimetically using supersaturated simulated body fluids on different substrates (Ti, Ti alloy, glass). DNDs were created by the detonation of carbon-containing explosives and the generated shockwave at high temperatures and pressures, which produced very fine particles of DND (approx. size: 4-6 nm). These DNDs were added to an SBF solution, from which the growth of an HAp-DND coating was stimulated on different substrates. Upon comparison, it was found that the composite coating was more compact than that of pure HAp but was porous. Cell culture testing revealed DND was not toxic to living cells. Similarly, Chen et al. [122] synthesised a biomedical HAp–nanodiamond based coating and deposited it using plasma spraying on a titanium substrate. The nanosized HAp powder was synthesised using the wet chemical route, with the powder being subsequently added in 3.0 wt.% PVA to make the slurry. The ND suspension was made separately using commercial ND particles and a solvent (a 1:1 water/ethanol mixture). The coating was achieved via spark plasma sintering using a 0.5 wt.% and 2.0 wt.% ND suspension. The coating proved to be of uniform structure with low porosity and had enhanced mechanical properties when compared to the equivalent pure HAp-based coatings.

Similarly, Pecheva et al. [123] incorporated ND particles into SBF solution to make a coating on a titanium substrate using an electrodeposition method for which they used a three-electrode electrolytic cell. ND particles were made via a shock-wave propagation method and were added into an SBF solution. The coating obtained was found to be of a homogenous structure, free from residual stresses and possessing high hardness and ductility values. Li et al. [124] reported a novel approach in which NDs were first biofunctionalized by attaching bone morphogenetic protein 2 to them and then made a composite with HAp. Detonation NDs for this study were made using a shock wave propagation method and were then converted into a suspension (NDS-PBS (phosphate buffered saline)). The composite powder was coated using a vacuum cold spray method.

3.3.6. Fullerenes

Fullerenes and their derivatives are found to have varied biomedical applications mostly in drug delivery systems. Fullerenes were discovered in 1985 and possess a carbon cage structure with immense scope for chemical derivatisation [125]. Fullerenes are entirely composed of carbon in the form of hollow spheres (or can be ellipsoidal in nature) or adopt related geometries. Highly symmetrical spherical fullerenes with icosahedral symmetry are known as Buckyballs or C_{60} [126]. Fullerenes are usually characterised by their hydrophobicity, three-dimensionality, electronic configuration, and good biological properties. The major issues related to the usage of fullerenes are its insolubility in aqueous media and aggregation [127]. One strategy to avoid such issues is chemical modification after which fullerene can be converted into fullerenol (a polyhydroxylated structure, $C_{60}(OH)_x$). This type of chemical modification is based on the addition of hydroxyl groups to the fullerene, which causes it to form into loosely associated, amorphous nano-aggregates. Djordjevic et al. [128] synthesised a hydroxyapatite/fullerenol nanocomposite via sonochemical processing; however, no mechanical characterisation was performed to access the strengthening effect. It was observed that the fullerenol nanoparticles had modified the surface of hydroxyapatite as the zeta potential values of the nanocomposite were ten times lower when compared to pure hydroxyapatite. This was attributed to the possibility of developing hydrogen bonding between surface phosphate groups in hydroxyapatite and hydroxyl groups in fullerenol.

3.3.7. Carbon Nanofibres

Carbon nanofibres (CNFs) are linear sp^2 carbon-based short fibres, having an aspect ratio greater than 100. They are generally vapour-grown or PAN (Polyacrylonitrile) based. CNFs possess excellent mechanical properties, biocompatibility, and non-toxicity, which render them suitable for reinforcement purposes in biomedical materials. Satoshi et al. [129] reported an increase of the mechanical properties in HAp based biocomposites due to carbon nanofibres (CNFs) reinforcement. The CNFs/HA composite was made via ball-milling techniques, followed by sintering using hot-pressing. It was observed that the addition of 10 vol.% CNFs resulted in a bending strength (i.e., 90 MPa), which was in the range of that of cortical bone. It was also found that the fracture toughness of CNFs/HAp composites was ca. 1.6 times higher than that of microporous HAp, but having the same level of bioactivity. Wu et al. [130]

developed a carbon nanofibre (CNF)/hydroxyapatite (HAp) composite, having strong interfacial bonding and high mechanical strength. The CNF was made using carbonisation of electrospun polymeric (polyacrylonitrile) precursor nanofibres. CNF mat was alkali-treated to introduce carboxylic groups onto its surface, and was then immersed in SBF solution to precipitate HAp on its surface. The fracture strength of 41% CNF-reinforced composites was found to be 67.3 MPa.

4. Other Systems Involving Carbon Structures and Hydroxyapatite

Researchers in the past have explored the strategy of reinforcing hydroxyapatite with suitable carbonaceous structures; however, there are numerous studies in which some other components have been added into the composite for biological and/or mechanical enhancement of the composite.

Herkendell et al. [131] highlighted the issue of bacterial infections during bone surgeries and emphasized the addition of a bactericidal component in the biomaterial to deal with the issue. They developed a CNT-reinforced HAp composite containing small amounts of silver (Ag), which is known for its antibacterial properties. A HAp–4 wt.% CNT–10 wt.% Ag composite demonstrated high density and a maximum fracture toughness enhancement of up to 244%. The composite was found to have good anti-bacterial properties together with high interfacial strength. With the addition of different trace metallic elements, researchers have found significant improvements in the biological properties of the HAp-based bioceramics. These have exploited the fact that the HAp crystal lattice is labile to these cations [132–134]. Different bioactive metallic ions have been investigated with HAp based composites, like Sr^{2+}, Mg^{2+}, Mn^{2+}, Ag^+, Zn^{2+}, and Y^{3+}. Among these, for instance, Mn^{2+} was found to influence the osteoblast differentiation and bone resorption and to promote biological activity and the ability to promote bone. Similarly, Zn^{2+} was found to stimulate new bone formation, increasing bone density, inhibiting osteoclastic proliferation and bone resorption in vivo. Elements incorporated into HAp, such as Ag^+ ions, need to be considered with some caution, however, because the presence of silver in vivo may lead to a medical condition known as Argyria in which permanent deposition of silver can occur in tissues causing cosmetic disfigurement if it is subcutaneous. This is discussed in detail in the paper by Hadrup and Iam [135].

Chen et al. [136] developed a chitosan- CNT-HAp nanocomposite using an in-situ precipitation method. Chitosan is a fibre extracted from chitin, which develops in a crustacean's exoskeleton. Chitosan is biocompatible and is biomedically exploited for various applications, such as bone tissue engineering [137–140] and drug delivery technology [141–143]. Multiwalled CNTs of 10 nm diameter were utilized and added into chitosan using ultrasonication. The composite was found to have good biocompatibility and also a maximum increase of ca. 110% in elastic modulus and 210% in compressive strength as the CNTs/chitosan wt. ratios rise from 0 to 5%. Yoon et al. [144] developed a gelatin-functionalised CNTs-HAp composite to mimic the natural structure of collagen fibrils in natural bone. In natural bone, collagen fibrils are interdigitated with HAp crystals and are responsible for strength and flexibility. Gelatine is a biocompatible, biodegradable, and high molecular weight polypeptide, which is already in medical usage, like wound dressings. Initially, in this reported research, a CNT/gelatin hybrid was made by covalently grafting gelatin molecules onto the surface of CNTs via the formation of amide linkages, then HAp crystals were assembled onto the CNT/gelatin hybridised surface. The structure of this composite is composed of CNT as a core and gelatin-HA as multi-layered shells. This hybrid material was found to be biocompatible and with enhanced mechanical properties.

Kalmodia et al. [145] manufactured a HAp -Alumina (Al_2O_3) -multiwalled carbon nanotubes (CNTs) composite. Previous, separate studies showed an enhancement of fracture toughness of the HAp matrix using individual Al_2O_3 reinforcement [146,147] and CNT reinforcement [72,93,98,99]. In this study, researchers were interested in investigating the combined reinforcing effect of CNT and Al_2O_3 in a HAp matrix. The composite was made while using spark plasma sintering and it was found to be biocompatible and also demonstrated favourable cell adhesion and proliferation properties. The highest hardness was recorded in a HAp-Al_2O_3 sample, but enhanced fracture toughness was observed when CNTs were added. Khanal et al. [72] utilised nylon along with carboxyl-functionalized CNTS (single walled) in a HAp matrix to enhance fracture toughness. The process comprised precipitation of HAp particles, which were then mixed with predefined quantities of CNTs and nylon. The fracture toughness values of the 1 wt.% carboxyl-functionalised single-walled CNTs and nylon were found to be 3.60 MPa.m$^{1/2}$.

Murugan et al. [148] found an enhancement in the biological and mechanical properties of mineralised hydroxyapatite reinforced by GO and oxidised CNF. GO was made using Hummer's method, while commercial CNF was treated while using NaOH. Among the different percentages trialled, 1% oxidised CNF and 1% GO-based composite were found to have the best properties for orthopaedic applications, like mechanical strength, which was 468 ± 4 HV (by Vicker's micro-hardness), along with favourable bactericidal properties.

5. Biomimetic Role of Hydroxyapatite and Carbon Composite Materials Intended as Biomaterials

Gustave Eiffel designed The Eiffel Tower after being inspired by the load distribution features from observing the "trabeculae" of human thigh bone (the longest and strongest bone in the body). In the biomedical field, using knowledge that is gained from biomimetics and biomineralisation can lead to biomaterials that are closer in function to natural boney tissues [149]. For an ideal implant material that can provide a natural environment for surrounding tissues and cells, it is necessary to meet the biochemical requirements of bone tissue engineering and to have the necessary mechanical properties to provide a framework for surrounding cells/tissue to interact with its surface. The environment leading to new tissue growth should eventually lead toward bone remodelling and new bone formation [16,150–152]. Designers for implant material(s) need to consider several different factors, which can either be extrinsic or intrinsic (Figure 8), however most of the extrinsic factors are outlined by medical professionals. An implant material is anticipated to complete the process of bone regeneration and healing after which implant removal becomes desirable from a clinical and biomechanical point of view. So, in such situations, bioactive and biodegradable materials are required to avoid a second surgery to remove them. Bioactive and biodegradable ceramics possess the unique ability of developing a bond to bone tissue, making them useful as 3D scaffolds and coatings [16,151,152]. Some of the most researched bioactive ceramics for bone augmentation are hydroxyapatite [34,153], β-tricalcium phosphate (β-TCP) [154,155], and bioactive glass [156–158]. They possess different rates and extents of resorption, however, all of them are biocompatible and osteoconductive.

Hydroxyapatite is found to have low biodegradability due to its low solubility (Ksp $\approx 6.62 \times 10^{-126}$) and it develops a bone-like calcium phosphate coating on the implant surface, which helps in bonding to the surrounding tissues [159,160]. Kim et al. [161] demonstrated how electrostatic interactions of the HAp surface, with Ca^{2+} and PO_4^{3-} ions promotes bone like apatite formation. The study was carried out using SBF. When an HAp surface is exposed to SBF, the surface displays negative charge and interacts with the Ca^{2+} ions, thereby forming a Ca-rich amorphous calcium phosphate layer, which resulted in the alteration of charge i.e., now the surface become positively charged. The process proceeds with the electrostatic interaction with PO_4^{3-} ions and forms the Ca-deficient amorphous calcium phosphate, which ultimately matures into bone-like apatite. Brandt et al. [162] reported hydroxyapatite implantation in rabbit femoral bone for 12 weeks, in which it was found that HAp exhibited bone formation, minimal degradation, and slow resorption. Following implantation, serum proteins are known to be readily adsorbed on the implant surface, which favourably alters the interfacial properties of the scaffold leading to initial in vivo resorption mostly from cellular activity. Tissue remodelling (bone formation and resorption) is carried out by osteoblast (bone forming) and osteoclast (bone resorbing) cells. When a hydroxyapatite-based implant is inserted into the physiological environment, a foreign body response is initiated, which follows a series of different phenomena, namely: injury (surgeons initially have to "injure" certain tissues and remove them to some extent to fix implant), protein matter interaction, blood clotting, inflammatory responses, fibrous tissue formation, and tissue remodelling. The resorption of HAp in the physiological environment creates space for newly developing tissues, which not only grows along the surface of the implant, but also infiltrates into the scaffold. This infiltration is accompanied with blood vessels (and oxygen supply for regenerating tissues), which eventually allows bone formation to progress. The initial degradation of HAp-based implants is highly dependent upon its properties, like porosity, surface roughness, and the site of implantation, as they directly affect the phenomena occurring near the implant surface (fluid exchange and nature of surrounding medium) [152]. Moreover, one of the most important parameters for bone augmentation is the volume of new bone formed, which is highly dependent on an adequate blood supply. Foreign body response near the implant results in a different physiological environment as there is found to be enhanced concentrations of reactive oxygen, proteolytic enzymes, fibrotic proteins, giant cells, as well as a reduced pH in the vicinity of the implant. The osteoinductive properties of bone-like materials are largely dependent upon calcium (Ca^{2+}) and phosphate (PO_4^{3-}) ions, as during in vivo bone resorption, osteoclasts resorb calcium (Ca^{2+}) and phosphate (PO_4^{3-}) ions from bone matrix, which results in a local increase of ion concentration. This ionic gradient assists in bone formation by the proliferation and differentiation of osteoblast cells. Then, the presence of interconnected macro and micro porosity (and its distribution) in HAp-based implants plays a significant role in boney tissue ingrowth, as it is responsible for oxygen and blood exchange during cell and bone growth [160].

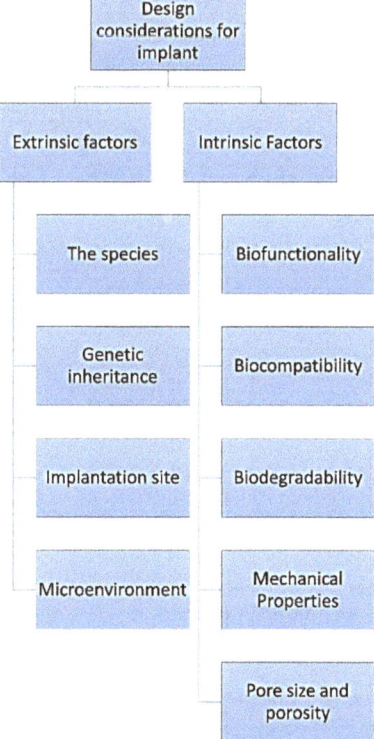

Figure 8. A general classification of several factors needs to be considered when designing an implant material.

For designing an effective biologically active surface in an implant material that can orchestrate these physiological healing process, the strong effect of altered environment near the implant is considered. In general, cellular response, which includes cell adhesion, proliferation, migration, and differentiation, is heavily influenced by the implant's surface chemistry, surface finish (the degree of roughness), topography, and wettability. There are several mechanisms responsible for effective cell adhesion to a biomaterial surface, including weak van der Waal's forces and electrostatic attraction; mechanical clinching to surface topographical structures and specific interactions between cell surface receptors and specific ligand molecules, as different cells react differently to the various surface features [16,151,163]. Jarcho [164] reported after clinical studies that HAp could develop direct physiological bonding to bone, with good biocompatibility and no inflammatory response. Initially, the surface undergoes partial solubility, which leads to direct deposition of calcified bone matrix. In vitro studies have shown that the presence of micro and macro porosities in HAp influence the degradation behaviour of HAp, but not the bonding of bone tissue. Wenisch et al. [165] studied the degradation of hydroxyapatite implanted in a sheep for six weeks, via transmission electron microscopy. The degradation was found to be osteoclast-mediated. Guda et al. [166] conducted a clinical animal study to find the effect of pore size on bone regeneration and reported that large uniform pore sizes result in enhanced bone regeneration.

Similarly, in the case of carbonaceous matter included in biomedical materials, their properties are mostly related to their surface features. Rajzer et al. [167] reported microscopic studies both in vivo and in vitro on two types of carbon fibres, in which one was HAp-modified, while the other type was a porous carbonaceous material. It was found out that cell adhesion was more effective on porous fibres when compared to the HAp-modified carbon fibres, which is most likely due to the increased surface area. Similarly, the significant amount of porosity also enhanced the tissue and cell growth within the implant structure, which has been observed to be the case with porous mineralized structures. As far as the biocompatibility of carbonaceous structures, especially carbon fibres, is concerned, it is known to be well tolerated by the body without any foreign reaction. More interestingly, the newly developed bony structure on the carbon scaffolds was found to be morphologically and functionally similar in form to the replaced natural bone. However, it was observed that in the case of fibrous carbon the cellular responses were inversely dependent upon the crystallinity i.e., amorphous fibres were judged to be excellent for implants rather that high crystallinity ones. Some studies [168,169] have revealed the adequate haemocompatibility; reduced thrombus formation in amorphous carbon. Du et al. [170] reported osteoblast cells attachment and proliferation on the surface of diamond like carbon (amorphous carbon) that was deposited on silicon substrates. Perkin and Naderi [171] reviewed and declared carbon nanostructures safe for clinical uses. They observed that, if the carbon structure is fixed in a solid composite structure, the effect of toxicity is reduced. Moreover, if the nanostructure detaches from the solid composite, then functionalisation will prevent its bio-accumulation. Schipper et al. [172] reported a clinical study in which functionalized-SWCNTs were injected into the bloodstream of mice. No toxic effect was noted for four months following treatment. Usui et al. [173] reported bone regeneration properties of MWCNTs, along with high bone tissue compatibility, as stimulated by human bone protein. Blazewicz [174] reported that highly carbonized (more crystalline) carbon fibres showed gradual fragmentation in biological environment as compared to low carbonised carbon fibres (i.e., with smaller crystallite sizes), which were easily resorbed and integrated in the body. Mild toxicity was also reported for CNTs mainly due to oxidative stress and inflammation [175]. Peterson [176] conducted an animal study on epoxy/carbon fibre-composite and reported that carbon fibres can undergo osseointegration with live bone.

There is a continued interest in making biocomposites with exceptional bone bonding abilities, for which both the scaffold matrix and its reinforcement are selected very prudently. The essential property that is required from a biocomposite is to demonstrate the formation of bone-like apatite on its surface. For designers, biocompatibility is not the only important parameter to consider for an ideally functioning implant material. It is also the strength, distribution of reinforcement, surface chemistry, porosity, microstructure, and surface roughness that comprise the other considerations related to the implant material. When a biocomposite is made using hydroxyapatite and carbonaceous structures (mainly for the purpose of overcoming the strength issues of hydroxyapatite), it is required that the resulting composite should also maintain adequate bone bonding properties, as otherwise the purpose of using a biologically inspired implant is incomplete. Hence the role of the reinforcing carbon is to maintain or minimally disrupt these bioactive properties of hydroxyapatite. As reported earlier, researchers have added different carbonaceous structures into hydroxyapatite, like nanocarbon forms, diamond and each resulting composite were found to possess different properties. In general, it is agreed that the carbon is bio-inert and has no toxic effect on living tissues [177]. Researchers have also evaluated the combined effects of HAp and carbon, either in the form of solid composites or as coatings. Mostly, nanostructures of carbon with HAp were evaluated for their biological properties. Khalid et al. [178] studied the interaction of functionalized-MWCNTs (f-MWCNTs) and f-MWCNTs-reinforced HAp composite with human osteoblast (sarcoma cell lines) in vitro. No damaging effect was noted on the survival of osteoblast cells, whereas it was noted that an increasing concentration of functionalized-MWCNTs had a cytotoxic effect. In a similar in vitro study, Khalid and Suman [179] reported no detrimental effect of f-MWCNTs HAp nanocomposites on a mouse fibroblast cell line. Liu et al. [180] demonstrated enhanced biocompatibility and cell growth properties by modifying pure carbon nanofibres using calcium phosphate. Similarly, Huang et al. [163] developed homogenous nanostructured calcium phosphate coatings on chemically modified carbon fibres using a biomineralisation process, which had promising biological properties. Han et al. [181] enhanced cell attachment and proliferation by modifying carbon fabric with calcium phosphate. Chlopek et al. [182] reported a modification of a carbon-carbon composite using hydroxyapatite powder. Upon testing them in an artificial biological environment, it was observed that the composite with a HAp-enriched surface provided favourable conditions for bone apatite growth. Newman et al. [183] developed a high porosity mechanically strong composite while using β-tricalcium phosphate/hydroxyapatite. A high-quality coating that is based on CNTs was applied to the composite for biomedical applications. Martinelli et al. [184] reported promising bone regeneration properties of nano HAP-CNT thin films developed on biomedical stainless steel via an in vitro osteogenesis study. Murugan, Murugan and Sundramoorthy [185] developed a biological coating based on hydroxyapatite/polycaprolactone-graphene oxide on Ti alloy. In vitro testing of this coating revealed outstanding cell viability, while in vivo testing in a rat model revealed bone formation after 28 days of implantation.

6. Conclusions and Future Aspects

In this review, the reinforcing effects of different carbonaceous structures in a HAp matrix (as scaffolds or coatings for orthopaedic applications) were discussed. The current demand on bone tissue engineering is to produce a scaffold or coating that should by itself be strong enough to bear loads, and to also possess favourable biocompatibility and bioactivity properties. This objective can be analysed in two parts as the first one deals with the mechanical properties, while the other deals with the biomedical/biocompatibility properties and out of the two, the maintenance of good biomedical properties is of the utmost importance i.e., it is crucial to select a material having exceptional biomedical properties, and then it is necessary to evaluate its overall strength and toughness. To produce such a robust material, scientists mostly rely on apatites/hydroxyapatite because of the highly favourable biomedical properties of these ceramics and then need to take steps to enhance its strength and toughness. However, reinforcement to enhance mechanical properties must be done without any compromise to the biological properties. Out of the many different materials, which can enhance the mechanical properties of HAp matrix, carbonaceous structures are always preferred in the biomedical science community because of their strength and biocompatibility. In this paper, it was highlighted that different structures, quantities, and modifications of carbon can be useful in increasing the strength and toughness of the HAp matrix. The biodegradability, osteoconductivity, and interface features of implant materials were also discussed, which suggests that the strategy to reinforce carbonaceous structure within the matrix of HAp will not disturb the bone regeneration properties of the composite.

As for future prospects, it is apparent that previous researchers have been focussed on achieving the maximum strength in a HAp matrix with use of the minimum amount of reinforcing the material, for which nano-reinforcements or nanofillers are a suitable choice. Moreover, it can also be observed that the morphology of reinforcing materials is also very important as in the case of nano carbonaceous materials where fibrous and sheet-like morphologies are found to be of value. When considering this, it can be suggested that future studies should exploit the combined effect of different carbonaceous materials. An interesting and important aspect deduced from previous work is that the full exploitation of HAp-carbonaceous structures-based composites has still not been achieved yet. This is because modifications in both the HAp and carbonaceous components, like functionalisation and mineralisation, still need to be considered to a greater extent to further improve the properties of the resulting composites. Finally, it is also recommended to utilise other carbonaceous structures, like carbon nanobuds (its structure is a like a hybrid of fullerene and CNT and it appears that a fullerene-like bud is attached to the wall of CNT), carbon nanoform, Q-carbon (non-crystalline carbon, having mixed sp^2 and sp^3 bonding), carbyne (long chains of carbon), and lonsdaleite (hexagonal diamond). In this way, a new generation of biomaterials can be produced, which greatly improves on previous ones.

Author Contributions: H.A.S. wrote the article with K.L.P. and M.R.M. providing research supervision, scientific/engineering feedback and editing suggestions for the article content and wording.

Funding: This research received no external funding.

Acknowledgments: The authors acknowledge the University of Waikato for all the essential resources/facilities for this work, especially Humair Ahmed Siddiqui who gladly acknowledges the University of Waikato for a University of Waikato Doctoral Scholarship. Humair Ahmed Siddiqui would also like to acknowledge NED University of Engg. and Tech. for facilitating higher studies.

Conflicts of Interest: The authors declare no conflicts of interest.

References

1. Chan, G.K.; Duque, G. Age-related bone loss: Old bone, new facts. *Gerontology* **2002**, *48*, 62–71. [CrossRef] [PubMed]
2. Swetha, M.; Sahithi, K.; Moorthi, A.; Srinivasan, N.; Ramasamy, K.; Selvamurugan, N. Biocomposites containing natural polymers and hydroxyapatite for bone tissue engineering. *Int. J. Biol. Macromol.* **2010**, *47*, 1–4. [CrossRef] [PubMed]
3. Amini, A.R.; Laurencin, C.T.; Nukavarapu, S.P. Bone Tissue Engineering: Recent Advances and Challenges. *Crit. Rev. Biomed. Eng.* **2012**, *40*, 363–408. [CrossRef] [PubMed]
4. Kanhed, S.; Awasthi, S.; Goel, S.; Pandey, A.; Sharma, R.; Upadhyaya, A.; Balani, K. Porosity distribution affecting mechanical and biological behaviour of hydroxyapatite bioceramic composites. *Ceram. Int.* **2017**, *43*, 10442–10449. [CrossRef]
5. Mauffrey, C.; Madsen, M.; Bowles, R.J.; Seligson, D. Bone graft harvest site options in orthopaedic trauma: A prospective in vivo quantification study. *Injury* **2012**, *43*, 323–326. [CrossRef] [PubMed]
6. Gao, C.; Feng, P.; Peng, S.; Shuai, C. Carbon nanotube, graphene and boron nitride nanotube reinforced bioactive ceramics for bone repair. *Acta Biomater.* **2017**, *61*, 1–20. [CrossRef] [PubMed]
7. Patel, N.R.; Gohil, P.P. A Review on Biomaterials: Scope, Applications & Human Anatomy Significance. *Int. J. Emerg. Technol. Adv. Eng.* **2012**, *2*, 91–101.
8. Wang, L.J.; Nancollas, G.H. Calcium Orthophosphates: Crystallization and Dissolution. *Chemical Reviews.* **2008**, *108*, 4628–4669. [CrossRef] [PubMed]
9. Dorozhkin, S.V. Calcium Orthophosphates in Nature, Biology and Medicine. *Materials* **2009**, *2*, 399–498. [CrossRef]
10. Dorozhkin, S.V. Bioceramics of calcium orthophosphates. *Biomaterials* **2010**, *31*, 1465–1485. [CrossRef] [PubMed]
11. LeGeros, R.Z. Calcium Phosphate-Based Osteoinductive Materials. *Chem. Rev.* **2008**, *108*, 4742–4753. [CrossRef] [PubMed]

12. Meyers, M.A.; Chen, P.Y.; Lin, A.Y.M.; Seki, Y. Biological materials: Structure and mechanical properties. *Prog. Mater. Sci.* **2008**, *53*, 1–206. [CrossRef]
13. Hench, L. Biomaterials. *Science* **1980**, *208*, 826–831. [CrossRef] [PubMed]
14. Hench, L.L.; Thompson, I. Twenty-first century challenges for biomaterials. *J. R. Soc. Interface* **2010**, *7* (Suppl. 4), S379–S391. [CrossRef]
15. Tan, L.; Yu, X.; Wan, P.; Yang, K. Biodegradable Materials for Bone Repairs: A Review. *J. Mater. Sci. Technol.* **2013**, *29*, 503–513. [CrossRef]
16. Navarro, M.; Michiardi, A.; Castaño, O.; Planell, J.A. Biomaterials in orthopaedics. *J. R. Soc. Interface* **2008**, *5*, 1137–1158. [CrossRef] [PubMed]
17. Stevens, M.M. Biomaterials for bone tissue engineering. *Mater. Today* **2008**, *11*, 18–25. [CrossRef]
18. O'Brien, F.J. Biomaterials & scaffolds for tissue engineering. *Mater. Today* **2011**, *14*, 88–95.
19. Taichman, R.S. Blood and bone: Two tissues whose fates are intertwined to create the hematopoietic stem-cell niche. *Blood* **2005**, *105*, 2631–2639. [CrossRef] [PubMed]
20. Suchanek, W.; Yoshimura, M. Processing and properties of hydroxyapatite-based biomaterials for use as hard tissue replacement implants. *J. Mater. Res.* **1998**, *13*, 94–117. [CrossRef]
21. Farraro, K.F.; Kim, K.E.; Woo, S.L.; Flowers, J.R.; McCullough, M.B. Revolutionizing orthopaedic biomaterials: The potential of biodegradable and bioresorbable magnesium-based materials for functional tissue engineering. *J. Biomech.* **2014**, *47*, 1979–1986. [CrossRef] [PubMed]
22. Mucalo, M.R. 14—Animal-bone derived hydroxyapatite in biomedical applications. In *Hydroxyapatite (HAp) for Biomedical Applications*; Woodhead Publishing: Cambridge, England, 2015; pp. 307–342.
23. Aoki, H. *Science and Medical Applications of Hydroxyapatite*; Japanese Association of Apatite Science: Tokyo, Japan, 1991.
24. Orlovskii, V.P.; Komlev, V.S.; Barinov, S.M. Hydroxyapatite and Hydroxyapatite-Based Ceramics. *Inorg. Mater.* **2002**, *38*, 973–984. [CrossRef]
25. Kraay, M.J.; Goldberg, V.M. Hydroxyapatite. In *Surgical Techniques in Total Knee Arthroplasty*; Scuderi, G.R., Tria, A.J., Jr., Eds.; Springer-Verlag: New York, NY, USA, 2002; pp. 277–286.
26. Shackelford, J.F. *Bioceramics*; Taylor & Francis: Boca Raton, FL, USA, 2003.
27. Basu, B.; Katti, D.S.; Kumar, A. *Advanced Biomaterials: Fundamentals, Processing, and Applications*; Wiley: Hoboken, NJ, USA, 2010.
28. Fihri, A.; Len, C.; Varma, R.S.; Solhy, A. Hydroxyapatite: A review of syntheses, structure and applications in heterogeneous catalysis. *Coord. Chem. Rev.* **2017**, *347*, 48–76. [CrossRef]
29. Rey, C.; Combes, C.; Drouet, C.; Glimcher, M.J. Bone mineral: Update on chemical composition and structure. *Osteoporos. Int.* **2009**, *20*, 1013–1021. [CrossRef] [PubMed]
30. Ana, I.D.; Matsuya, S.; Ishikawa, K. Engineering of Carbonate Apatite Bone Substitute Based on Composition-Transformation of Gypsum and Calcium Hydroxide. *Engineering* **2010**, *2*, 344. [CrossRef]
31. Hench, L.L. *An Introduction to Bioceramics: Second Edition*; World Scientific Publishing Company: Singapore, 2013.
32. Paul, W.; Sharma, C. Development of porous spherical hydroxyapatite granules: Application towards protein delivery. *J. Mater. Sci. Mater. Med.* **1999**, *10*, 383–388.
33. Weinlander, M.; Plenk, H., Jr.; Adar, F.; Holmes, R. TCP- Impurities in HA- Granules and Crystallinity Changes in Plasmaflamesprayed HA- Coatings Detected by Spectroscopical Methods and their Consequence. In *Bioceramics and the Human Body*; Ravaglioli, A., Krajewski, A., Eds.; Elsevier Science Publishers: New York, NY, USA, 1992.
34. Zhou, H.; Lee, J. Nanoscale hydroxyapatite particles for bone tissue engineering. *Acta Biomater.* **2011**, *7*, 2769–2781. [CrossRef] [PubMed]
35. Hing, K.; Best, S.; Tanner, K.; Bonfield, W.; Revell, P. Quantification of bone ingrowth within bone-derived porous hydroxyapatite implants of varying density. *J. Mater. Sci. Mater. Med.* **1999**, *10*, 663–670. [CrossRef] [PubMed]
36. Yamamoto, M.; Tabata, Y.; Kawasaki, H.; Ikada, Y. Promotion of fibrovascular tissue ingrowth into porous sponges by basic fibroblast growth factor. *J. Mater. Sci. Mater. Med.* **2000**, *11*, 213–218. [CrossRef] [PubMed]

37. Johnson, G.; Mucalo, M.; Lorier, M. The processing and characterization of animal-derived bone to yield materials with biomedical applications Part 1: Modifiable porous implants from bovine condyle cancellous bone and characterization of bone materials as a function of processing. *J. Mater. Sci. Mater. Med.* **2000**, *11*, 427–441. [CrossRef] [PubMed]
38. Worth, A.; Mucalo, M.; Horne, G.; Bruce, W.; Burbidge, H. The evaluation of processed cancellous bovine bone as a bone graft substitute. *Clin. Oral. Implants Res.* **2005**, *16*, 379–386. [CrossRef] [PubMed]
39. Callan, D.P.; Rohrer, M.D. Use of bovine-derived hydroxyapatite in the treatment of edentulous ridge defects: A human clinical and histologic case report. *J. Periodontol.* **1993**, *64*, 575–582. [CrossRef] [PubMed]
40. Yukna, R.A.; Callan, D.P.; Krauser, J.T.; Evans, G.H.; Aichelmann-Reidy, M.E.; Moore, K.; Cruz, R.; Scott, J.B. Multi-center clinical evaluation of combination anorganic bovine-derived hydroxyapatite matrix (ABM)/cell binding peptide (P-15) as a bone replacement graft material in human periodontal osseous defects. 6-month results. *J. Periodontol.* **1998**, *69*, 655–663. [CrossRef] [PubMed]
41. Boutinguiza, M.; Pou, J.; Comesaña, R.; Lusquiños, F.; de Carlos, A.; León, B. Biological hydroxyapatite obtained from fish bones. *Mater. Sci. Eng. C* **2012**, *32*, 478–486. [CrossRef]
42. Cahyanto, A.; Kosasih, E.; Aripin, D.; Hasratiningsih, Z. Fabrication of hydroxyapatite from fish bones waste using reflux method. *IOP Conf. Ser. Mater. Sci. Eng.* **2017**, *172*, 012006. [CrossRef]
43. Sunil, B.R.; Jagannatham, M. Producing hydroxyapatite from fish bones by heat treatment. *Mater. Lett.* **2016**, *185*, 411–414. [CrossRef]
44. Raya, I.; Mayasari, E.; Yahya, A.; Syahrul, M.; Latunra, A.I. Shynthesis and Characterizations of Calcium Hydroxyapatite Derived from Crabs Shells (Portunus pelagicus) and Its Potency in Safeguard against to Dental Demineralizations. *Int. J. Biomater.* **2015**, *2015*, 469176. [CrossRef] [PubMed]
45. Lee, S.-W.; Balázsi, C.; Balázsi, K.; Seo, D.-h.; Kim, H.S.; Kim, C.-H.; Kim, S.-G. Comparative Study of hydroxyapatite prepared from seashells and eggshells as a bone graft material. *Tissue Eng. Regen. Med.* **2014**, *11*, 113–120. [CrossRef]
46. Vecchio, K.S.; Zhang, X.; Massie, J.B.; Wang, M.; Kim, C.W. Conversion of bulk seashells to biocompatible hydroxyapatite for bone implants. *Acta Biomater.* **2007**, *3*, 910–918. [CrossRef] [PubMed]
47. Kattimani, V.S.; Chakravarthi, P.S.; Kanumuru, N.R.; Subbarao, V.V.; Sidharthan, A.; Kumar, T.S.S.; Prasad, L.K. Eggshell Derived Hydroxyapatite as Bone Graft Substitute in the Healing of Maxillary Cystic Bone Defects: A Preliminary Report. *J. Int. Oral Health* **2014**, *6*, 15–19. [PubMed]
48. Wu, S.-C.; Hsu, H.-C.; Hsu, S.-K.; Chang, Y.-C.; Ho, W.-F. Synthesis of hydroxyapatite from eggshell powders through ball milling and heat treatment. *J. Asian Ceram. Soc.* **2016**, *4*, 85–90. [CrossRef]
49. Rivera, E.M.; Araiza, M.; Brostow, W.; Castaño, V.M.; Díaz-Estrada, J.R.; Hernández, R.; Rodríguez, J.R. Synthesis of hydroxyapatite from eggshells. *Mater. Lett.* **1999**, *41*, 128–134. [CrossRef]
50. Yunus Basha, R.; Kumar, T.S.S.; Doble, M. Design of biocomposite materials for bone tissue regeneration. *Mater. Sci. Eng. C Mater. Biol. Appl.* **2015**, *57*, 452–463. [CrossRef] [PubMed]
51. John, M. and S. Thomas, Biofibres and biocomposites. *Carbohydr. Polym.* **2008**, *71*, 343–364. [CrossRef]
52. Launey, M.E.; Ritchie, R.O. On the Fracture Toughness of Advanced Materials. *Adv. Mater.* **2009**, *21*, 2103–2110. [CrossRef]
53. Ritchie, R.O. The conflicts between strength and toughness. *Nat. Mater.* **2011**, *10*, 817. [CrossRef] [PubMed]
54. Anderson, T.L. *Fracture Mechanics: Fundamentals and Applications*, 3rd ed.; Taylor & Francis: Boca Raton, FL, USA, 2005.
55. Gogotsi, G.A. Fracture toughness of ceramics and ceramic composites. *Ceram. Int.* **2003**, *29*, 777–784. [CrossRef]
56. Lemons, J.E. Ceramics: Past, present, and future. *Bone* **1996**, *19*, 121s–128s. [CrossRef]
57. Yin, H.; Qi, H.J.; Fan, F.; Zhu, T.; Wang, B.; Wei, Y. Griffith Criterion for Brittle Fracture in Graphene. *Nano Lett.* **2015**, *15*, 1918–1924. [CrossRef] [PubMed]
58. Sitarz, M.; Drajewicz, M.; Jadach, R.; Długoń, E.; Leśniak, M.; Reben, M.; Wajda, A.; Gawęda, M.; Burtan-Gwizdała, B. Optical and Mechanical Characterization of Zirconium Based Sol-Gel Coatings on Glass. *Arch. Metall. Mater.* **2016**, *61*, 1747–1752. [CrossRef]
59. Ohji, T.; Jeong, Y.-K.; Choa, Y.-H.; Niihara, K. Strengthening and Toughening Mechanisms of Ceramic Nanocomposites. *J. Am. Ceram. Soc.* **1998**, *81*, 1453–1460. [CrossRef]
60. Wang, G.C.; Lu, Z.F.; Zreiqat, H. 8—Bioceramics for skeletal bone regeneration A2—Mallick, Kajal. In *Bone Substitute Biomaterials*; Woodhead Publishing: Cambridge, England, 2014; pp. 180–216.

61. Zhao, G.; Huang, C.; Liu, H.; Zou, B.; Zhu, H.; Wang, J. A study on in-situ synthesis of TiB2–SiC ceramic composites by reactive hot pressing. *Ceram. Int.* **2014**, *40 Pt B*, 2305–2313. [CrossRef]
62. Naslain, R.R.; Khan, T.; Steen, M.; Holmes, P. Ceramic Matrix Composites [and Discussion]. *Philos. Trans. Phys. Sci. Eng.* **1995**, *351*, 485–496.
63. Hirsch, A. The era of carbon allotropes. *Nat. Mater.* **2010**, *9*, 868. [CrossRef] [PubMed]
64. Zahra, K.; Majid, M.; Mircea, V.D. Main Allotropes of Carbon: A Brief Review. In *Sustainable Nanosystems Development, Properties, and Applications*; Mihai, V.P., Marius Constantin, M., Eds.; IGI Global: Hershey, PA, USA, 2017; pp. 185–213.
65. Falcao Eduardo, H.L.; Wudl, F. Carbon allotropes: Beyond graphite and diamond. *J. Chem. Technol. Biotechnol.* **2007**, *82*, 524–531. [CrossRef]
66. Bharath, K.N.; Basavarajappa, S. Applications of biocomposite materials based on natural fibers from renewable resources: A review. *Sci. Eng. Compos. Mater.* **2016**, *23*, 123–133. [CrossRef]
67. Fan, X.; Case, E.D.; Ren, F.; Shu, Y.; Baumann, M.J. Part I: Porosity dependence of the Weibull modulus for hydroxyapatite and other brittle materials. *J. Mech. Behav. Biomed. Mater.* **2012**, *8*, 21–36. [CrossRef] [PubMed]
68. Nahorny, S.; Zanin, H.; Christino, V.A.; Marciano, F.R.; Lobo, A.O.; Soares, L.E.S. Multi-walled carbon nanotubes/graphene oxide hybrid and nanohydroxyapatite composite: A novel coating to prevent dentin erosion. *Mater. Sci. Eng. C Mater. Biol. Appl.* **2017**, *79*, 199–208. [CrossRef] [PubMed]
69. Georgakilas, V.; Perman, J.A.; Tucek, J.; Zboril, R. Broad Family of Carbon Nanoallotropes: Classification, Chemistry, and Applications of Fullerenes, Carbon Dots, Nanotubes, Graphene, Nanodiamonds, and Combined Superstructures. *Chem. Rev.* **2015**, *115*, 4744–4822. [CrossRef] [PubMed]
70. Awaji, H.; Choi, S.-M.; Yagi, E. Mechanisms of toughening and strengthening in ceramic-based nanocomposites. *Mech. Mater.* **2002**, *34*, 411–422. [CrossRef]
71. Sekhar, S.; Bal, S. Carbon Nanotube Reinforced Ceramic Matrix Composites—A Review. *J. Miner. Mater. Charact. Eng.* **2008**, *7*, 355–370.
72. Khanal, S.P.; Mahfuz, H.; Rondinone, A.J.; Leventouri, T. Improvement of the fracture toughness of hydroxyapatite (HAp) by incorporation of carboxyl functionalized single walled carbon nanotubes (CfSWCNTs) and nylon. *Mater. Sci. Eng. C* **2016**, *60*, 204–210. [CrossRef] [PubMed]
73. Gao, L.; Lin, J. Electrophoretic coating of Hydroxyapatite on pyrolytic carbon using glycol as dispersion medium. *J. Wuhan Univ. Technol. Mater. Sci. Ed.* **2008**, *23*, 293–297. [CrossRef]
74. Ratner, B.D.; Hoffman, A.S.; Schoen, F.J.; Lemons, J.E. *Biomaterials Science: An Introduction to Materials in Medicine*, 3rd ed.; Academic Press: Cambridge, MA, USA, 2012.
75. Hetherington, V.J.; Lord, C.E.; Brown, S.A. Mechanical and histological fixation of hydroxylapatite-coated pyrolytic carbon and titanium alloy implants: A report of short-term results. *J. Appl. Biomater.* **1995**, *6*, 243–248. [CrossRef] [PubMed]
76. Fitzer, E.; Gkogkidis, A.; Heine, M. Carbon Fibers and their Composites (A Review). *High. Temp.-High. Press.* **1984**, *16*, 363–392.
77. Zakharov, N.A.; Safonova, A.M.; Orlov, M.A.; Demina, L.I.; Aliev, A.D.; Kiselev, M.R.; Matveev, V.V.; Shelekhov, E.V.; Zakharova, T.V.; Kuznetsov, N.T. Synthesis and properties of calcium hydroxyapatite/carbon fiber composites. *Russ. J. Inorg. Chem.* **2017**, *62*, 1162–1172. [CrossRef]
78. Ślósarczyk, A.; Klisch, M.; Błażewicz, M.; Piekarczyk, J.; Stobierski, L.; Rapacz-Kmita, A. Hot Pressed Hydroxyapatite–Carbon Fibre Composites. *J. Eur. Ceram. Soc.* **2000**, *20*, 1397–1402. [CrossRef]
79. Dorner-Reisel, A.; Berroth, K.; Neubauer, R.; Nestler, K.; Marx, G.; Scislo, M.; Müller, E.; Slosarcyk, A. Unreinforced and carbon fiber reinforced hydroxyapatite: Resistance against microabrasion. *J. Eur. Ceram. Soc.* **2004**, *24*, 2131–2139. [CrossRef]
80. Boehm, A.; Meininger, S.; Tesch, A.; Gbureck, U.; Müller, F. The Mechanical Properties of Biocompatible Apatite Bone Cement Reinforced with Chemically Activated Carbon Fibers. *Materials* **2018**, *11*, 192. [CrossRef] [PubMed]
81. Kuo, T.F.; Chi, C.C.; Lin, I.N. Synthesis of Carbon Nanotubes by Laser Ablation of Graphites at Room Temperature. *Jpn. J. Appl. Phys.* **2001**, *40*, 7147. [CrossRef]
82. Akasaka, T.; Watari, F.; Sato, Y.; Tohji, K. Apatite formation on carbon nanotubes. *Mater. Sci. Eng. C* **2006**, *26*, 675–678. [CrossRef]

83. Aryal, S.; Bhattarai, S.R.; KC, R.B.; Khil, M.S.; Lee, D.-R.; Kim, H.Y. Carbon nanotubes assisted biomimetic synthesis of hydroxyapatite from simulated body fluid. *Mater. Sci. Eng. A* **2006**, *426*, 202–207. [CrossRef]
84. Arora, N.; Sharma, N.N. Arc discharge synthesis of carbon nanotubes: Comprehensive review. *Diam. Relat. Mater.* **2014**, *50*, 135–150. [CrossRef]
85. Sharma, R.; Sharma, A.K.; Sharma, V. Synthesis of carbon nanotubes by arc-discharge and chemical vapor deposition method with analysis of its morphology, dispersion and functionalization characteristics. *Cogent Eng.* **2015**, *2*, 1094017. [CrossRef]
86. Arepalli, S. Laser ablation process for single-walled carbon nanotube production. *J. Nanosci. Nanotechnol.* **2004**, *4*, 317–325. [CrossRef] [PubMed]
87. Chrzanowska, J.; Hoffman, J.; Małolepszy, A.; Mazurkiewicz, M.; Kowalewski, T.A.; Szymanski, Z.; Stobinski, L. Synthesis of carbon nanotubes by the laser ablation method: Effect of laser wavelength. *Phys. Status Solidi* **2015**, *252*, 1860–1867. [CrossRef]
88. Brukh, R.; Mitra, S. Mechanism of carbon nanotube growth by CVD. *Chem. Phys. Lett.* **2006**, *424*, 126–132. [CrossRef]
89. Kumar, M.; Ando, Y. Chemical vapor deposition of carbon nanotubes: A review on growth mechanism and mass production. *J. Nanosci. Nanotechnol.* **2010**, *10*, 3739–3758. [CrossRef] [PubMed]
90. Chen, Y.; Gan, C.; Zhang, T.; Yu, G.; Bai, P.; Kaplan, A. Laser-surface-alloyed carbon nanotubes reinforced hydroxyapatite composite coatings. *Appl. Phys. Lett.* **2005**, *86*, 251905. [CrossRef]
91. Lahiri, D.; Singh, V.; Keshri, A.K.; Seal, S.; Agarwal, A. Carbon nanotube toughened hydroxyapatite by spark plasma sintering: Microstructural evolution and multiscale tribological properties. *Carbon* **2010**, *48*, 3103–3120. [CrossRef]
92. Lei, T.; Wang, L.; Ouyang, C.; Li, N.-F.; Zhou, L.-S. In Situ Preparation and Enhanced Mechanical Properties of Carbon Nanotube/Hydroxyapatite Composites. *Int. J. Appl. Ceram. Technol.* **2011**, *8*, 532–539. [CrossRef]
93. Kealley, C.; Ben-Nissan, B.; van Riessen, A.; Elcombe, M. Development of Carbon Nanotube Reinforced Hydroxyapatite Bioceramics. *Key Eng. Mater.* **2006**, *309*, 597–602. [CrossRef]
94. Kaya, C.; Singh, I.; Boccaccini, A.R. Multi-walled Carbon Nanotube-Reinforced Hydroxyapatite Layers on Ti6Al4V Medical Implants by Electrophoretic Deposition (EPD). *Adv. Eng. Mater.* **2008**, *10*, 131–138. [CrossRef]
95. Ye, M.; Wenjiang, Q.; Jingqin, P. Preparation and characterization of mechanical properties of carbon nanotube reinforced hydroxyapatite composites consolidated by spark plasma sintering. *IOP Conf. Ser. Mater. Sci. Eng.* **2017**, *231*, 012164.
96. Sarkar, S.; Ho Youn, M.; Hyun Oh, I.; Taek Lee, B. Fabrication of CNT-Reinforced HAp Composites by Spark Plasma Sintering. *Mater. Sci. Forum* **2007**, *534–536*, 893–896. [CrossRef]
97. Balani, K.; Anderson, R.; Laha, T.; Andara, M.; Tercero, J.; Crumpler, E.; Agarwal, A. Plasma-sprayed carbon nanotube reinforced hydroxyapatite coatings and their interaction with human osteoblasts in vitro. *Biomaterials* **2007**, *28*, 618–624. [CrossRef] [PubMed]
98. Li, H.; Zhao, Q.; Li, B.; Kang, J.; Yu, Z.; Li, Y.; Song, X.; Liang, C.; Wang, H. Fabrication and properties of carbon nanotube-reinforced hydroxyapatite composites by a double in situ synthesis process. *Carbon* **2016**, *101*, 159–167. [CrossRef]
99. Kim, D.-Y.; Han, Y.-H.; Lee, J.H.; Kang, I.-K.; Jang, B.-K.; Kim, S. Characterization of Multiwalled Carbon Nanotube-Reinforced Hydroxyapatite Composites Consolidated by Spark Plasma Sintering. *BioMed Res. Int.* **2014**, *2014*, 10. [CrossRef] [PubMed]
100. Lee, H.C.; Liu, W.-W.; Chai, S.-P.; Mohamed, A.R.; Lai, C.W.; Khe, C.-S.; Voon, C.H.; Hashim, U.; Hidayah, N.M.S. Synthesis of Single-layer Graphene: A Review of Recent Development. *Procedia Chem.* **2016**, *19*, 916–921. [CrossRef]
101. Allen, M.J.; Tung, V.C.; Kaner, R.B. Honeycomb Carbon: A Review of Graphene. *Chem. Rev.* **2010**, *110*, 132–145. [CrossRef] [PubMed]
102. Ghuge, A.D.; Shirode, A.R.; Kadam, V.J. Graphene: A Comprehensive Review. *Curr. Drug Targets* **2017**, *18*, 724–733. [CrossRef] [PubMed]
103. Liu, Y.; Huang, J.; Li, H. Synthesis of hydroxyapatite-reduced graphite oxide nanocomposites for biomedical applications: Oriented nucleation and epitaxial growth of hydroxyapatite. *J. Mater. Chem. B* **2013**, *1*, 1826–1834. [CrossRef]

104. Zhang, L.; Liu, W.; Yue, C.; Zhang, T.; Li, P.; Xing, Z.; Chen, Y. A tough graphene nanosheet/hydroxyapatite composite with improved in vitro biocompatibility. *Carbon* **2013**, *61*, 105–115. [CrossRef]
105. Gurunathan, S.; Kim, J.-H. Synthesis, toxicity, biocompatibility, and biomedical applications of graphene and graphene-related materials. *Int. J. Nanomed.* **2016**, *11*, 1927–1945. [CrossRef] [PubMed]
106. Chen, D.; Feng, H.; Li, J. Graphene Oxide: Preparation, Functionalization, and Electrochemical Applications. *Chem. Rev.* **2012**, *112*, 6027–6053. [CrossRef] [PubMed]
107. Zaaba, N.I.; Foo, K.L.; Hashim, U.; Tan, S.J.; Liu, W.-W.; Voon, C.H. Synthesis of Graphene Oxide using Modified Hummers Method: Solvent Influence. *Procedia Eng.* **2017**, *184*, 469–477. [CrossRef]
108. Pei, S.; Cheng, H.-M. The reduction of graphene oxide. *Carbon* **2012**, *50*, 3210–3228. [CrossRef]
109. Schafhaeutl, C. LXXXVI. On the combinations of carbon with silicon and iron, and other metals, forming the different species of cast iron, steel, and malleable iron. *Lond. Edinb. Dublin Philos. Mag. J. Sci.* **1840**, *16*, 570–590. [CrossRef]
110. Brodie, B.C. XIII. On the atomic weight of graphite. *Philos. Trans. R. Soc. Lond.* **1859**, *149*, 249–259. [CrossRef]
111. Hummers, W.S.; Offeman, R.E. Preparation of Graphitic Oxide. *J. Am. Chem. Soc.* **1958**, *80*, 1339. [CrossRef]
112. Li, M.; Liu, Q.; Jia, Z.; Xu, X.; Cheng, Y.; Zheng, Y.; Xi, T.; Wei, S. Graphene oxide/hydroxyapatite composite coatings fabricated by electrophoretic nanotechnology for biological applications. *Carbon* **2014**, *67*, 185–197. [CrossRef]
113. Fathyunes, L.; Khalil-Allafi, J. Characterization and corrosion behavior of graphene oxide-hydroxyapatite composite coating applied by ultrasound-assisted pulse electrodeposition. *Ceram. Int.* **2017**, *43*, 13885–13894. [CrossRef]
114. Li, Y.; Liu, C.; Zhai, H.; Zhu, G.; Pan, H.; Xu, X.; Tang, R. Biomimetic graphene oxide–hydroxyapatite composites via in situ mineralization and hierarchical assembly. *RSC Adv.* **2014**, *4*, 25398–25403. [CrossRef]
115. Abdolhosseinzadeh, S.; Asgharzadeh, H.; Kim, H.S. Fast and fully-scalable synthesis of reduced graphene oxide. *Sci. Rep.* **2015**, *5*, 10160. [CrossRef] [PubMed]
116. Baradaran, S.; Moghaddam, E.; Basirun, W.J.; Mehrali, M.; Sookhakian, M.; Hamdi, M.; Moghaddam, M.R.N.; Alias, Y. Mechanical properties and biomedical applications of a nanotube hydroxyapatite-reduced graphene oxide composite. *Carbon* **2014**, *69*, 32–45. [CrossRef]
117. Öztürk, E.; Özbek, B.; Şenel, İ. Production of biologically safe and mechanically improved reduced graphene oxide/hydroxyapatite composites. *Mater. Res. Express* **2017**, *4*, 015601.
118. Mengesha, A.E.; Youan, B.B.C. 8—Nanodiamonds for drug delivery systems. In *Diamond-Based Materials for Biomedical Applications*; Narayan, R., Ed.; Woodhead Publishing: Cambridge, England, 2013; pp. 186–205.
119. Ansari, S.A.; Satar, R.; Jafri, M.A.; Rasool, M.; Ahmad, W.; Kashif Zaidi, S. Role of Nanodiamonds in Drug Delivery and Stem Cell Therapy. *Iran. J. Biotechnol.* **2016**, *14*, 130–141. [CrossRef] [PubMed]
120. Mochalin, V.N.; Shenderova, O.; Ho, D.; Gogotsi, Y. The properties and applications of nanodiamonds. *Nat. Nanotechnol.* **2011**, *7*, 11. [CrossRef] [PubMed]
121. Pramatarova, L.; Pecheva, E.; Dimitrova, R.; Spassov, T.; Krasteva, N.; Hikov, T.; Fingarova, D.; Mitev, D. Hydroxyapatite Reinforced Coatings with Incorporated Detonationally Generated Nanodiamonds. *AIP Conf. Proc.* **2010**, *1203*, 937–942.
122. Chen, X.; Zhang, B.; Gong, Y.; Zhou, P.; Li, H. Mechanical properties of nanodiamond-reinforced hydroxyapatite composite coatings deposited by suspension plasma spraying. *Appl. Surf. Sci.* **2018**, *439*, 60–65. [CrossRef]
123. Pecheva, E.; Pramatarova, L.; Toth, A.; Hikov, T.; Fingarova, D.; Stavrev, S.; Iacob, E.; Vanzetti, L. Effect of Nanodiamond Particles Incorporation in Hydroxyapatite Coatings. *ECS Trans.* **2009**, *25*, 403–410.
124. Li, D.; Chen, X.; Gong, Y.; Zhang, B.; Liu, Y.; Jin, P.; Li, H. Synthesis and Vacuum Cold Spray Deposition of Biofunctionalized Nanodiamond/Hydroxyapatite Nanocomposite for Biomedical Applications. *Adv. Eng. Mater.* **2017**, *19*, 1700363. [CrossRef]
125. Geng, J.; Miyazawa, K.; Hu, Z.; Solov'yov, I.A.; Berenguer-Murcia, A. Fullerene-Related Nanocarbons and Their Applications. *J. Nanotechnol.* **2012**, *2012*, 2.
126. Taylor, R.; Hare, J.P.; Abdul-Sada, A.A.K.; Kroto, H.W. Isolation, separation and characterisation of the fullerenes C60 and C70: The third form of carbon. *J. Chem. Soc.Chem. Commun.* **1990**, 1423–1425. [CrossRef]
127. Prato, M. [60] Fullerene chemistry for materials science applications. *J. Mater. Chem.* **1997**, *7*, 1097–1109. [CrossRef]

128. Djordjevic, A.; Ignjatovic, N.; Seke, M.; Danica, J.; Uskoković, D.; Zlatko, R. Synthesis and Characterization of Hydroxyapatite/Fullerenol Nanocomposites. *J. Nanosci. Nanotechnol.* **2015**, *15*, 1538–1542. [CrossRef] [PubMed]
129. Kobayashi, S.; Kawai, W. Development of carbon nanofiber reinforced hydroxyapatite with enhanced mechanical properties. *Compos. Part A Appl. Sci. Manuf.* **2007**, *38*, 114–123. [CrossRef]
130. Wu, M.; Wang, Q.; Liu, X.; Liu, H. Biomimetic synthesis and characterization of carbon nanofiber/hydroxyapatite composite scaffolds. *Carbon* **2013**, *51*, 335–345. [CrossRef]
131. Herkendell, K.; Shukla, V.R.; Patel, A.K.; Balani, K. Domination of volumetric toughening by silver nanoparticles over interfacial strengthening of carbon nanotubes in bactericidal hydroxyapatite biocomposite. *Mater. Sci. Eng. C Mater. Biol. Appl.* **2014**, *34*, 455–467. [CrossRef] [PubMed]
132. Evis, Z.; Webster, T.J. Nanosize hydroxyapatite: Doping with various ions. *Adv. Appl. Ceram.* **2011**, *110*, 311–321. [CrossRef]
133. Sallam, S.M.; Tohami, K.M.; Sallam, A.M.; Salem, L.I.A.; Mohamed, F.A. Synthesis and characterization of hydroxyapatite contain chromium. *J. Biophys. Chem.* **2012**, *3*, 278. [CrossRef]
134. Zilm, M.E.; Chen, L.; Sharma, V.; McDannald, A.; Jain, M.; Ramprasad, R.; Wei, M. Hydroxyapatite substituted by transition metals: Experiment and theory. *Phys. Chem. Chem. Phys.* **2016**, *18*, 16457–16465. [CrossRef] [PubMed]
135. Hadrup, N.; Lam, H.R. Oral toxicity of silver ions, silver nanoparticles and colloidal silver—A review. *Regul. Toxicol. Pharmacol.* **2014**, *68*, 1–7. [CrossRef] [PubMed]
136. Chen, L.; Hu, J.; Shen, X.; Tong, H. Synthesis and characterization of chitosan-multiwalled carbon nanotubes/hydroxyapatite nanocomposites for bone tissue engineering. *J. Mater. Sci. Mater. Med.* **2013**, *24*, 1843–1851. [CrossRef] [PubMed]
137. Venkatesan, J.; Kim, S.-K. Chitosan Composites for Bone Tissue Engineering—An Overview. *Mar. Drugs* **2010**, *8*, 2252–2266. [CrossRef] [PubMed]
138. Levengood, S.L.; Zhang, M. Chitosan-based scaffolds for bone tissue engineering. *J. Mater. Chem. B Mater. Boil. Med.* **2014**, *2*, 3161–3184. [CrossRef] [PubMed]
139. LogithKumar, R.; KeshavNarayan, A.; Dhivya, S.; Chawla, A.; Saravanan, S.; Selvamurugan, N. A review of chitosan and its derivatives in bone tissue engineering. *Carbohydr. Polym.* **2016**, *151*, 172–188. [CrossRef] [PubMed]
140. Saravanan, S.; Leena, R.S.; Selvamurugan, N. Chitosan based biocomposite scaffolds for bone tissue engineering. *Int. J. Biol. Macromol.* **2016**, *93*, 1354–1365. [CrossRef] [PubMed]
141. Wang, J.J.; Zeng, Z.W.; Xiao, R.Z.; Xie, T.; Zhou, G.L.; Zhan, X.R.; Wang, S.L. Recent advances of chitosan nanoparticles as drug carriers. *Int. J. Nanomed.* **2011**, *6*, 765–774.
142. Bernkop-Schnürch, A.; Dünnhaupt, S. Chitosan-based drug delivery systems. *Eur. J. Pharm. Biopharm.* **2012**, *81*, 463–469. [CrossRef] [PubMed]
143. Paul, W.; Sharma, C.P. Chitosan, a drug carrier for the 21st century: A review. *STP Pharma Sci.* **2000**, *10*, 5–22.
144. Yoon, I.-K.; Hwang, J.-Y.; Seo, J.-w.; Jang, W.-C.; Kim, H.-W.; Shin, U.S. Carbon nanotube-gelatin-hydroxyapatite nanohybrids with multilayer core–shell structure for mimicking natural bone. *Carbon* **2014**, *77*, 379–389. [CrossRef]
145. Kalmodia, S.; Goenka, S.; Laha, T.; Lahiri, D.; Basu, B.; Balani, K. Microstructure, mechanical properties, and in vitro biocompatibility of spark plasma sintered hydroxyapatite–aluminum oxide–carbon nanotube composite. *Mater. Sci. Eng. C* **2010**, *30*, 1162–1169. [CrossRef]
146. Sivaperumal, V.R.; Mani, R.; Nachiappan, M.S.; Arumugam, K. Direct hydrothermal synthesis of hydroxyapatite/alumina nanocomposite. *Mater. Charact.* **2017**, *134*, 416–421. [CrossRef]
147. Horng Yih, J.; Hsiung, H.M. Fabrication and mechanical properties of hydroxyapatite-alumina composites. *Mater. Sci. Eng. C* **1994**, *2*, 77–81. [CrossRef]
148. Murugan, N.; Anandhakumar, S.; Shen-Ming, C.; Ashok, K.S. Graphene oxide/oxidized carbon nanofiber/ mineralized hydroxyapatite based hybrid composite for biomedical applications. *Mater. Rese. Express* **2017**, *4*, 124005. [CrossRef]
149. Sailaja, G.S.; Ramesh, P.; Vellappally, S.; Anil, S.; Varma, H.K. Biomimetic approaches with smart interfaces for bone regeneration. *J. Biomed. Sci.* **2016**, *23*, 77. [CrossRef] [PubMed]
150. Wu, T.; Yu, S.; Chen, D.; Wang, Y. Bionic Design, Materials and Performance of Bone Tissue Scaffolds. *Materials* **2017**, *10*, 1187. [CrossRef] [PubMed]

151. Prakasam, M.; Locs, J.; Salma-Ancane, K.; Loca, D.; Largeteau, A.; Berzina-Cimdina, L. Biodegradable Materials and Metallic Implants—A Review. *J. Funct. Biomater.* **2017**, *8*, 44. [CrossRef] [PubMed]
152. Kokubo, T. *Bioceramics and their Clinical Applications*; Woodhead Publishing: Cambridge, England, 2008.
153. Oonishi, H.; Yamamoto, M.; Ishimaru, H.; Tsuji, E.; Kushitani, S.; Aono, M.; Ukon, Y. The effect of hydroxyapatite coating on bone growth into porous titanium alloy implants. *J. Bone Joint Surg.* **1989**, *71*, 213–216. [CrossRef]
154. LeGeros, R.Z. Properties of Osteoconductive Biomaterials: Calcium Phosphates. *Clin. Orthop. Relat. Res.* **2002**, *395*, 81–98. [CrossRef]
155. Daculsi, G.; Laboux, O.; Malard, O.; Weiss, P. Current state of the art of biphasic calcium phosphate bioceramics. *J. Mater. Sci. Mater. Med.* **2003**, *14*, 195–200. [CrossRef] [PubMed]
156. Rizwan, M.; Hamdi, M.; Basirun, W.J. Bioglass(R) 45S5-based composites for bone tissue engineering and functional applications. *J. Biomed. Mater. Res. A* **2017**, *105*, 3197–3223. [CrossRef] [PubMed]
157. Hench, L.L.; Paschall, H.A. Direct chemical bond of bioactive glass-ceramic materials to bone and muscle. *J. Biomed. Mater. Res.* **1973**, *7*, 25–42. [CrossRef] [PubMed]
158. Oonishi, H.; Hench, L.L.; Wilson, J.; Sugihara, F.; Tsuji, E.; Matsuura, M.; Kin, S.; Yamamoto, T.; Mizokawa, S. Quantitative comparison of bone growth behavior in granules of Bioglass®, A-W glass-ceramic, and hydroxyapatite. *J. Biomed. Mater. Res.* **2000**, *51*, 37–46. [CrossRef]
159. Raucci, M.G.; Guarino, V.; Ambrosio, L. Biomimetic Strategies for Bone Repair and Regeneration. *J. Funct. Biomater.* **2012**, *3*, 688–705. [CrossRef] [PubMed]
160. Tang, Z.; Li, X.; Tan, Y.; Fan, H.; Zhang, X. The material and biological characteristics of osteoinductive calcium phosphate ceramics. *Regen. Biomater.* **2018**, *5*, 43–59. [CrossRef] [PubMed]
161. Kim, H.M.; Himeno, T.; Kawashita, M.; Kokubo, T.; Nakamura, T. The mechanism of biomineralization of bone-like apatite on synthetic hydroxyapatite: An in vitro assessment. *J. R. Soc. Interface* **2004**, *1*, 17–22. [CrossRef] [PubMed]
162. Brandt, J.; Henning, S.; Michler, G.; Hein, W.; Bernstein, A.; Schulz, M. Nanocrystalline hydroxyapatite for bone repair: An animal study. *J. Mater. Sci. Mater. Med.* **2009**, *21*, 283–294. [CrossRef] [PubMed]
163. Huang, S.-P.; Zhou, K.-C.; Li, Z.-Y. Preparation and mechanism of calcium phosphate coatings on chemical modified carbon fibers by biomineralization. *Trans. Nonferr. Metals Soc. China* **2008**, *18*, 162–166. [CrossRef]
164. Jarcho, M. Calcium phosphate ceramics as hard tissue prosthetics. *Clin. Orthop. Relat. Res.* **1981**, *157*, 259–278.
165. Wenisch, S.; Stahl, J.P.; Horas, U.; Heiss, C.; Kilian, O.; Trinkaus, K.; Hild, A.; Schnettler, R. In vivo mechanisms of hydroxyapatite ceramic degradation by osteoclasts: Fine structural microscopy. *J. Biomed. Mater. Res. A* **2003**, *67*, 713–718. [CrossRef] [PubMed]
166. Guda, T.; Walker, J.A.; Singleton, B.; Hernandez, J.; Oh, D.S.; Appleford, M.R.; Ong, J.L.; Wenke, J.C. Hydroxyapatite scaffold pore architecture effects in large bone defects in vivo. *J. Biomater. Appl.* **2014**, *28*, 1016–1027. [CrossRef] [PubMed]
167. Rajzer, I.; Menaszek, E.; Bacakova, L.; Rom, M.; Blazewicz, M. In vitro and in vivo studies on biocompatibility of carbon fibres. *J. Mater. Sci. Mater. Med.* **2010**, *21*, 2611–2622. [CrossRef] [PubMed]
168. Jones, M.I.; McColl, I.R.; Grant, D.M.; Parker, K.G.; Parker, T.L. Haemocompatibility of DLC and TiC–TiN interlayers on titanium. *Diam. Relat. Mater.* **1999**, *8*, 457–462. [CrossRef]
169. Chen, J.Y.; Wang, L.; Fu, K.Y.; Huang, l.; Leng, Y.X.; Yang, P.; Wang, J.; Wan, G.J.; Sun, H.; Tian, X.B.; Chu, P.K. Blood compatibility and sp3/sp2 contents of diamond-like carbon (DLC) synthesized by plasma immersion ion implantation-deposition. *Surf. Coat. Technol.* **2002**, *156*, 289–294.
170. Du, C.; Su, X.W.; Cui, F.Z.; Zhu, X.D. Morphological behaviour of osteoblasts on diamond-like carbon coating and amorphous C–N film in organ culture. *Biomaterials* **1998**, *19*, 651–658. [CrossRef]
171. Perkins, B.L.; Naderi, N. Carbon Nanostructures in Bone Tissue Engineering. *Open Orthop. J.* **2016**, *10*, 877–899. [CrossRef] [PubMed]
172. Schipper, M.L.; Nakayama-Ratchford, N.; Davis, C.R.; Kam, N.W.; Chu, P.; Liu, Z.; Sun, X.; Dai, H.; Gambhir, S.S. A pilot toxicology study of single-walled carbon nanotubes in a small sample of mice. *Nat. Nanotechnol.* **2008**, *3*, 216–221. [CrossRef] [PubMed]
173. Usui, Y.; Aoki, K.; Narita, N.; Murakami, N.; Nakamura, I.; Nakamura, K.; Ishigaki, N.; Yamazaki, H.; Horiuchi, H.; Kato, H.; Taruta, S.; Kim, Y.A.; Endo, M.; Saito, N. Carbon nanotubes with high bone-tissue compatibility and bone-formation acceleration effects. *Small* **2008**, *4*, 240–246. [CrossRef] [PubMed]

174. Blazewicz, M. Carbon materials in the treatment of soft and hard tissue injuries. *Eur. Cell Mater.* **2001**, *2*, 21–29. [CrossRef] [PubMed]
175. Simona, C.; Adriana, F. In vivo Assessment of Nanomaterials Toxicity. In *Nanomaterials—Toxicity and Risk Assessment*; Marcelo, L.L., Soloneski, S., Eds.; InTech: London, UK, 2015; pp. 93–121.
176. Petersen, R. Carbon Fiber Biocompatibility for Implants. *Fibers* **2016**, *4*, 1. [CrossRef] [PubMed]
177. Rodil, S.; Olivares-Navarrete, R.; Arzate, H.; Muhl, S. Biocompatibility, Cytotoxicity and Bioactivity of Amorphous Carbon Films. *Carbon* **2006**, *100*, 55–75.
178. Khalid, P.; Hussain, M.A.; Rekha, P.D.; Arun, A.B. Carbon nanotube-reinforced hydroxyapatite composite and their interaction with human osteoblast in vitro. *Hum. Exp. Toxicol.* **2015**, *34*, 548–556. [CrossRef] [PubMed]
179. Khalid, P.; Suman, V.B. Carbon nanotube-hydroxyapatite composite for bone tissue engineering and their interaction with mouse fibroblast L929 In Vitro. *J. Bionanosci.* **2017**, *11*, 233–240. [CrossRef]
180. Liu, H.; Cai, Q.; Lian, P.; Fang, Z.; Duan, S.; Ryu, S.; Yang, X.; Deng, X. The biological properties of carbon nanofibers decorated with β-tricalcium phosphate nanoparticles. *Carbon* **2010**, *48*, 2266–2272. [CrossRef]
181. Han, H.-M.; Phillips, G.J.; Mikhalovsky, S.V.; Lloyd, A.W. In vitro cytotoxicity assessment of carbon fabric coated with calcium phosphate. *New Carbon Mater.* **2008**, *23*, 139–143. [CrossRef]
182. Chlopek, J.; Blazewicz, M.; Szaraniec, B. Effect of artificial biological environment on mechanical properties of carbon-phosphate composites. *Kompozyty* **2002**, *2*, 163–166.
183. Newman, P.; Lu, Z.; Roohani-Esfahani, S.I.; Church, T.L.; Biro, M.; Davies, B.; King, A.; Mackenzie, K.; Minett, A.I.; Zreiqat, H. Porous and strong three-dimensional carbon nanotube coated ceramic scaffolds for tissue engineering. *J. Mater. Chem. B* **2015**, *3*, 8337–8347. [CrossRef]
184. Martinelli, N.M.; Ribeiro, M.J.G.; Ricci, R.; Marques, M.A.; Lobo, A.O.; Marciano, F.R. In Vitro Osteogenesis Stimulation via Nano-Hydroxyapatite/Carbon Nanotube Thin Films on Biomedical Stainless Steel. *Materials* **2018**, *11*, 1555.
185. Murugan, N.; Murugan, C.; Sundramoorthy, A.K. In vitro and in vivo characterization of mineralized hydroxyapatite/polycaprolactone-graphene oxide based bioactive multifunctional coating on Ti alloy for bone implant applications. *Arab. J. Chem.* **2018**, *11*, 959–969. [CrossRef]

© 2018 by the authors. Licensee MDPI, Basel, Switzerland. This article is an open access article distributed under the terms and conditions of the Creative Commons Attribution (CC BY) license (http://creativecommons.org/licenses/by/4.0/).

MDPI
St. Alban-Anlage 66
4052 Basel
Switzerland
Tel. +41 61 683 77 34
Fax +41 61 302 89 18
www.mdpi.com

Materials Editorial Office
E-mail: materials@mdpi.com
www.mdpi.com/journal/materials

www.ingramcontent.com/pod-product-compliance
Lightning Source LLC
LaVergne TN
LVHW071951080526
838202LV00064B/6724